Wine Aging Technologies

Wine Aging Technologies

Special Issue Editors

María Del Alamo-Sanza
Ignacio Nevares

MDPI • Basel • Beijing • Wuhan • Barcelona • Belgrade

Special Issue Editors

María Del Alamo-Sanza
Universidad de Valladolid
Spain

Ignacio Nevares
Universidad de Valladolid
Spain

Editorial Office
MDPI
St. Alban-Anlage 66
4052 Basel, Switzerland

This is a reprint of articles from the Special Issue published online in the open access journal *Beverages* (ISSN 2306-5710) from 2018 to 2019 (available at: https://www.mdpi.com/journal/beverages/special_issues/wine_aging_technologies)

For citation purposes, cite each article independently as indicated on the article page online and as indicated below:

LastName, A.A.; LastName, B.B.; LastName, C.C. Article Title. *Journal Name* **Year**, *Article Number*, Page Range.

ISBN 978-3-03897-748-3 (Pbk)
ISBN 978-3-03897-749-0 (PDF)

Cover image courtesy of María del Alamo-Sanza and Ignacio Nevares.

Contents

About the Special Issue Editors

María del Alamo-Sanza, Dr., has been an Associated Professor of the Department of Analytical Chemistry of the University of Valladolid (Spain) since 2007. She graduated with her doctorate in Chemistry from the University of Valladolid in 1997. Her research has always been related to wine chemistry, wine phenolics, oak wood, oxygen, and wine, as well as the characterization of wine aging in different aging systems (traditional and alternatives) and oak woods. She is the head and a cofounder of the UVaMOX research group focused on oxygen and wine, control of wine maturation processes, evaluation of the oxygen transfer rate in barrels, vessels made by different natural materials, stoppers, characterization of the silent micro-oxygenation in the cellar, and wine oxygen management.

Ignacio Nevares, Dr., is currently an Associated Professor at the Department of Agricultural and Forestry Engineering of the Higher Tech. Col. of Agricultural Engineering at the University of Valladolid (Spain). He received his degree as an Agricultural Engineering Specialist in Agricultural and Food Industries from the Polytechnic University of Madrid (Spain) in 1991 and his PhD in Agricultural Engineering from the University of Valladolid in 2003. He is a cofounder of the UVaMOX research group at the University of Valladolid, and he is a specialist on traditional wine aging as well as alternative aging techniques where oxygen plays an important role. He also works in winemaking automation. His main current interest is oxygen measurement and its role in wine aging processes, specifically toward improving the performance of micro-oxygenation devices, elucidating the behavior of the barrel, and optimizing oxygen and wine-related technical procedures.

Editorial

Wine Aging Technologies

Maria Del Alamo-Sanza [1,*] and **Ignacio Nevares** [2,*]

1 Department of Analytical Chemistry, UVaMOX-Higher Tech. Col. of Agricultural Engineering, Universidad de Valladolid, 34001 Palencia, Spain
2 Department of Agroforestry Engineering, UVaMOX-Higher Tech. Col. of Agricultural Engineering, Universidad de Valladolid, 34001 Palencia, Spain
* Correspondence: maria.alamo.sanza@uva.es (M.D.A.-S.); ignacio.nevares@uva.es (I.N.)

Received: 7 March 2019; Accepted: 7 March 2019; Published: 12 March 2019

Wine aging is a desirable and valuable process, commonly used to improve wine quality, and traditionally carried out in oak wooden casks. The correct use of oak barrels and the ever-increasing demand for barrels in different production areas of the world has led to a constant search for technological alternatives to reproduce the chemical and physical processes undergone by wines during their stay in barrels.

This Special Issue aims to publish a compilation of original research and reviews that cover different aspects of the aging processes of wine in casks and other alternative systems that reproduce, with different technologies, the transformations that take place in the barrel. This special issue has seven works, the first, titled "Different Woods in Cooperage for Oenology: A Review" [1] by Ana Martínez-Gil et al., focused on the possible use of different woods to the traditional ones in the wine aging process. New trends in the use of barrels have resulted in an increased demand for oak wood in the cooperage. This growing demand has led to the use of woods within the genus *Quercus* other than those traditionally used (*Quercus alba*, *Quercus petraea*, and *Quercus robur*) and even wood of different genera. The species of the genus *Quercus*, such as *Quercus pyrenaica* Willd, *Quercus faginea* Lam, *Quercus humboldtti* Bonpl, *Quercus oocarpa* Liebm, *Quercus frainetto* Ten, and other genera, such as *Robinia pseudoacacia* L. (false acacia), *Castanea sativa* Mill, *Prunus avium* L. and *Prunus cereaus* L. (cherry), *Fraxinus excelsior* L. (European ash), *Fraxinus americana* L. (American ash), *Morus nigra* L. and *Morus alba* L. have been studied as possible sources of wood suitable for cooperage. The chemical characterization of these woods is fundamental to be able to adapt the treatment of the cooperage and therefore obtain a wood with enological qualities suitable for the treatment of wines. This review aims to summarize the different species that have been studied as possible new sources of wood for enology, defining the extractable composition of each of them and their use in wine.

The second work entitled "Study of High Power Ultrasound for Oak Wood Barrel Regeneration: Impact on Wood Properties and Sanitation Effect" [2], by Breniauxa et al., presented the ability of high power ultrasound (HPU) to ensure oak barrel sterilization and wood structure preservation. Optimization was performed in terms of temperature and time. The impact of the HPU process on the porous material was also characterized. In this research, several wood characteristics were considered, such as the specific surface area, hydrophobicity, oxygen desorption, and spoilage microorganisms after treatment. The study showed that the microbial stabilization could be obtained with HPU 60 °C/6 min. The results obtained show that microorganisms are impacted up to a depth of 9 mm, with a *Brettanomyces bruxellensis* population <1 log CFU/g. The operating parameters used during the HPU treatment can also impact wood exchange surface and oxygen desorption kinetics indicating that tartrate was removed. Indeed, the total oxygen desorption rate was recovered after HPU treatment close to a new oak barrel and this may indicate that there is no impact on the ultrastructure (vessel, pore size, or rays). Finally, wood wettability can also be impacted depending on the temperature and the duration of exposure.

The third work was carried out within the scope of alternative aging systems, with the title "Oxygen Consumption by Red Wines under Different Micro-Oxygenation Strategies and *Q. Pyrenaica* Chips. Effects on Color and Phenolic Characteristics" [3] by Sánchez-Gómez et al. described the importance of micro-oxygenation (MOX) as a key factor in obtaining a final wine that is more stable over time and with similar characteristics as barrel-aged wines. Therefore, the oxygen dosage added must be that which the wine can consume to develop correctly. The oxygen consumption of red wine determines its properties so it is essential that micro-oxygenation is managed properly. This paper shows the results from the study of the influence on red wine of two different MOX strategies: floating oxygen dosage (with dissolved oxygen setpoint of 50 μg/L) and fixed oxygen dosage (3 mL/L ·month). The results indicated that the wines consumed all the oxygen provided: those from fixed MOX received between 3 and 3.5 times more oxygen than the floating MOX strategy; the oxygen contribution from the air entrapped in the wood was more significant in the latter. Wines aged with wood and MOX showed the same color and phenolic evolution as those aged in barrels, which demonstrates the importance of MOX management. Despite the differences in oxygen consumption, it was not possible to differentiate wines from the different MOX strategies at the end of the aging period. Another work in the same field is that of Rubio-Bretón et al. entitled "Use of Oak Fragments during the Aging of Red Wines. Effect on the Phenolic, Aromatic, and Sensory Composition of Wines as a Function of the Contact Time with the Wood [4]. This paper studied the effect of the use of oak fragments on the volatile, phenolic, and sensory characteristics of Tempranillo red wines, as a function of the contact time between the wood and the wine. The results showed important changes in the wines' colorimetric parameters after two months of contact time. Extraction kinetics of volatile compounds from the wood were highest during the first month of contact for chips, variable for staves, and slower and continuous over time for barrels. Wines macerated with fragments showed the best quality in short periods of aging, while barrel-aged wines improved over the time they spent in the barrel. In addition, the results allowed an analytical discrimination between the wines aged with oak fragments and those aged in oak barrels, and between chips and staves, just as at the sensory level with triangular taste tests. In conclusion, the use of oak fragments is a suitable practice for the production of red wines, which may be an appropriate option for wines destined to be aged for short periods. The fifth paper, "New Strategies to Improve Sensorial Quality of White Wines by Wood Contact" [5] by Alañón et al., focused on the importance of quantitative and qualitative compounds of the wood depending on the species, its origins, and the treatments applied in cooperages. Traditionally, oak wood species are most often used in cooperage, specifically *Quercus alba*, known as American oak, and *Quercus robur* and *Quercus petraea*, both known as French oak. Although the use of wood contact is very common for red wines, its use is still restricted in the case of white wines. However, this topic is particularly interesting, since due to the sensorial benefits of wood contact, the option for aging white wines in barrels or chips could be chosen by winemakers. This review compiles the novel strategies applied to white wines using wood contact in recent years with the aim to increase wine quality and sensorial features.

Mohekar et al. presented their work "Effects of Fining Agents, Reverse Osmosis and Wine Age on Brown Marmorated Stink Bug (*Halyomorpha halys*) Taint in Wine" [6], in which is they described how trans-2-decenal and tridecane are compounds found in wine made from brown marmorated stink bug (BMSB)-contaminated grapes. The effectiveness of the post-fermentation processes on reducing their concentration in finished wine and their longevity during wine aging was evaluated. Red wines containing trans-2-decenal were treated with fining agents and put through reverse osmosis filtration. The efficiency of these treatments was determined using chemical analysis (multidimensional gas chromatography–mass spectrometry (MDGC–MS)) and sensory descriptive analysis. Tridecane and trans-2-decenal concentrations in red and white wine were determined at bottle aging durations of 0, 6, 12, and 24 months using MDGC–MS. Reverse osmosis was found to be partially successful in removing trans-2-decenal concentrations from finished wine. While tridecane and trans-2-decenal concentrations decreased during bottle aging, post-fermentative fining treatments were not effective at removing these compounds. Although French oak did not alter the concentration of tridecane and trans-2-decenal in red wine, it did mask the expression of BMSB-related sensory characters. Because of the ineffectiveness of removing BMSB taint post-fermentation, BMSB densities in the grape clusters should be minimized so that taints do not occur in the wine.

And finally the paper of Rodríguez et al., entitled "Novel Method for the Identification of the Variety of Grape Using Their Capability to Form Gold Nanoparticles" [7], showed the possibilities of gold nanoparticles (AuNPs) obtained using musts (freshly prepared grape juices where solid peels and seeds have been removed) as a reducing and capping agent. Transmission electron microscope images showed that the formed AuNPs were spherical and their size increased with the amount of must used. The size of the AuNPs increased with an increase in the total polyphenol index (TPI) of the variety of grape. The kinetics of the reaction monitored using UV-VIS showed that the reaction rates were related to the chemical composition of the musts and specifically to the phenols that acted as reducing and capping agents during the synthesis process. Since the particular composition of each must produces AuNPs of different sizes and at different rates, color changes can be used to discriminate the variety of grape. This new technology can be used to avoid fraud.

We would like to acknowledge the chance offered by MDPI, the publisher, to coordinate and serve as guest editors of this special issue of Beverages on the topic of wine aging technologies, which we have pleasantly carried out. We are very grateful to the authors who have generously shared their scientific knowledge and experience with others through their contribution to this special issue.

Conflicts of Interest: The authors declare no conflict of interest.

References

1. Martínez-Gil, A.; Del Alamo-Sanza, M.; Sánchez-Gómez, R.; Nevares, I. Different Woods in Cooperage for Oenology: A Review. *Beverages* **2018**, *4*, 94. [CrossRef]
2. Breniaux, M.; Renault, P.; Meunier, F.; Ghidossi, R. Study of High Power Ultrasound for Oak Wood Barrel Regeneration: Impact on Wood Properties and Sanitation Effect. *Beverages* **2019**, *5*, 10. [CrossRef]
3. Sánchez-Gómez, R.; Nevares, I.; Martínez-Gil, A.M.; Del Alamo-Sanza, M. Oxygen Consumption by Red Wines under Different Micro-Oxygenation Strategies and Q. Pyrenaica Chips. Effects on Color and Phenolic Characteristics. *Beverages* **2018**, *4*, 69. [CrossRef]
4. Rubio-Bretón, P.; Garde-Cerdán, T.; Martínez, J. Use of Oak Fragments during the Aging of Red Wines. Effect on the Phenolic, Aromatic, and Sensory Composition of Wines as a Function of the Contact Time with the Wood. *Beverages* **2018**, *4*, 102. [CrossRef]
5. Alañón, M.E.; Díaz-Maroto, M.C.; Pérez-Coello, M.S. New Strategies to Improve Sensorial Quality of White Wines by Wood Contact. *Beverages* **2018**, *4*, 91. [CrossRef]
6. Mohekar, P.; Osborne, J.; Tomasino, E. Effects of Fining Agents, Reverse Osmosis and Wine Age on Brown Marmorated Stink Bug (Halyomorpha halys) Taint in Wine. *Beverages* **2018**, *4*, 17. [CrossRef]
7. Rodriguez, S.; De Lamo, B.; García-Hernández, C.; García-Cabezón, C.; Rodríguez-Méndez, M.L. Novel Method for the Identification of the Variety of Grape Using Their Capability to Form Gold Nanoparticles. *Beverages* **2018**, *4*, 26. [CrossRef]

Review

Different Woods in Cooperage for Oenology: A Review

Ana Martínez-Gil [1], Maria del Alamo-Sanza [1], Rosario Sánchez-Gómez [1] and Ignacio Nevares [2,*]

1   Department of Analytical Chemistry, UVaMOX-Higher Tech. Col. of Agricultural Engineering, Universidad de Valladolid, 34001 Palencia, Spain; anamaria.martinez.gil@uva.es (A.M.-G.); maria.alamo.sanza@uva.es (M.d.A.-S.); rosario.sanchez@uva.es (R.S.-G.)
2   Department of Agroforestry Engineering, UVaMOX-Higher Tech. Col. of Agricultural Engineering, Universidad de Valladolid, 34001 Palencia, Spain
*   Correspondence: ignacio.nevares@uva.es

Received: 31 October 2018; Accepted: 19 November 2018; Published: 23 November 2018

Abstract: Contact of wine with wood during fermentation and ageing produces significant changes in its chemical composition and organoleptic properties, modifying its final quality. Wines acquire complex aromas from the wood, improve their colour stability, flavour, and clarification, and extend their storage period. New trends in the use of barrels, replaced after a few years of use, have led to an increased demand for oak wood in cooperage. In addition, the fact that the wine market is becoming increasingly saturated and more competitive means that oenologists are increasingly interested in tasting different types of wood to obtain wines that differ from those already on the market. This growing demand and the search for new opportunities to give wines a special personality has led to the use of woods within the *Quercus* genus that are different from those used traditionally (*Quercus alba*, *Quercus petraea*, and *Quercus robur*) and even woods of different genera. Thus, species of the genus *Quercus*, such as *Quercus pyrenaica* Willd., *Quercus faginea* Lam., *Quercus humboldtti* Bonpl., *Quercus oocarpa* Liebm., *Quercus frainetto* Ten, and other genera, such as *Robinia pseudoacacia* L. (false acacia), *Castanea sativa* Mill. (chestnut), *Prunus avium* L. and *Prunus cereaus* L. (cherry), *Fraxinus excelsior* L. (European ash), *Fraxinus americana* L. (American ash), *Morus nigra* L, and *Morus alba* L. have been the subject of several studies as possible sources of wood apt for cooperage. The chemical characterization of these woods is essential in order to be able to adapt the cooperage treatment and, thus, obtain wood with oenological qualities suitable for the treatment of wines. This review aims to summarize the different species that have been studied as possible new sources of wood for oenology, defining the extractable composition of each one and their use in wine.

Keywords: traditional oaks; different oaks; other woods; ellagitannins; low molecular phenols; volatile compounds

1. Introduction

The wine trade controlled mainly by Greeks and Romans (2000 BCE) used earthenware jars and amphora, although these containers were fragile, heavy, and difficult to handle. Faced with this problem of transporting wines from the production to the consumption areas, wooden containers were created. The study of archaeological findings and written testimonies allows us to establish how wooden barrels displaced clay amphorae for wine transport and storage: a revolution. There are many references to the use of wooden containers for wine. The best-known reference is possibly that of Julius Caesar in "The Gallic Wars" (51 BCE) [1,2]. From the 5th century onwards, the term 'barrel' was used to designate these wooden containers. Since then, oak has been one of the main woods for this purpose, being a resistant, flexible, easy to handle, and not very permeable material. Specifically, the

European species, mainly *Quercus petraea* and *Quercus robur*, were used as they were abundant near the areas where wine was made.

In the mid-twentieth century, the use of wood was notably abandoned due to the proliferation of other materials (cement and stainless steel). However, from the 1990s onwards, the use of wooden barrels re-emerged rather significantly and became a world fashion [2–4]. In addition, this resurgence also led to a change in their use. Nowadays, wine ageing has changed with the use of newer oak barrels because with their use the extractability of the oak compounds decreases. For this reason, in recent years, an imbalance between the amount of oak available and the number of barrels produced has been detected in France [5]. Moreover, the price of cooperage logs is increasing steadily with a concomitant decrease in the quality/price ratio. Given the growing demand for French oak barrels (*Quercus petraea* and *Quercus robur*) and the increase in price, some cooperages also work with oak from Eastern Europe (Romania, Hungary, Russia), as it is the same species with characteristics similar to those of French oak. Many oenologists have used barrels of this type of oak, as the results obtained are comparable to those of ageing wine in French oak. For these reasons, over the last few years, there has been a proliferation of studies on European oaks of the same species but of different origins. In recent years, the literature has offered studies about Slovenian oak [6], Spanish oaks [7–11], Hungarian oaks [12,13], Russian oaks [13,14], Romanian, Ukrainian, and Moldavian oaks [15], Romanian oak [16,17], among others. These studies of the same species but different origin showed similar characteristics to American and French oaks, suggesting that they are suitable for barrel production for quality wines. Some authors even state that these origins have intermediate characteristics between French and American oaks [7,9,15].

Singleton mentions different woods within cooperage, from both the United States (white oak, red oak, chestnut oak, red or sweet gum, sugar maple, yellow or sweet birch, white ash, Douglas fir, beech, black cherry, sycamore, redwood, spruce, bald cypress, elm, and basswood) and Europe (white oak, chestnut, fir, spruce, pine, larch, ash, mulberry), and a number of additional species imported from Africa, South America, and Australia (acacia, karri: *Eucalyptus diversicolor*; jarrah: *Eucalyptus marginata*; stringybark: *Eucalyptus obliqua* and *Eucalyptus gigantea*; and she oak: *Casuarina fraseriana*). However, this long list of woods rapidly diminishes when considering only those that are suitable for ageing different alcoholic beverages [18]. As the number of wood species declined, oak and chestnut became the most widely used varieties in barrel-making and so were already those most used from the 16th century onwards [19–21]. They were chosen because they modified the gustatory and olfactory characteristics of the different wines and spirits favourably [21]. These two woods (*Quercus* and *Chestnut*) are the only ones approved today by the Organzation of Vine and Wine (OIV) (Resolution OENO 4/2005).

The typical anatomy of oak offers greater resistance, flexibility, easy handling, and low permeability in relation to those provided by other woods [22]. At present, oak (*Quercus*) is the preferred material for the manufacture of barrels for ageing alcoholic beverages, especially wines. Oak belongs to the genus *Quercus*, which is made up of more than 250 species, although this figure is controversial as some authors cite up to 600 [21,23]. Most of these are to be found in the temperate zones of the northern hemisphere as far as south to Central America and Ecuador. The number of species increases from East to West, from Europe and Africa to the North American Pacific coast, Mexico being the country with the greatest diversity of species. The *Quercus* genus is subdivided into two subgenera, Cyclobalanopsis and Euquercus: the first includes tropical species and some from Asia and Malaysia not used in the manufacture of barrels for oenological use, while those within the subgenera Euquercus are used in cooperage [21]. Within these species, few meet all of the requirements, and those most used belong to the group of white oaks [5]. According to Vivas [21], some of the species used in cooperage in the USA and Europe are *Quercus alba*, *Quercus garryana*, *Quercus macrocarpa*, and *Quercus stellata*, and only in Europe *Quercus cerris*, *Quercus suber*, *Quercus coccifera*, *Quercus lanuginosa*, *Quercus petraea*, and *Quercus robur*. The main species used for wine ageing belong to the genus *Oersted* (formerly Lepidobalanus): *Quercus petraea* L. (*Quercus sessilis*) and *Quercus robur* (*Quercus pedunculata*) growing in Europe and *Quercus alba* growing in different areas of the United States.

In America, white oak has only been associated with *Q. Alba* for years; however, strictly speaking, the classification of "white oak" includes many other species, such as: *Q. alba*, *Q. garryana*, *Q. macrocarpa* M., *Q. stellata* Wan., *Quercus lyrata* Walt., *Quercus prinus*, *Quercus muehlenbergii* E., *Quercus michauxi* Nutt, *Quercus bicolor* Willd., *Quercus lobata* Née, *Quercus montana* Willd, and *Quercus virginiana* L. [24,25]. Thus, in cooperage, *Q. alba*, a majority species in the eastern United States, has been associated with numerous species resulting in confusion, as is the case for: *Q. prinus*, *Q. muehlenbergii*, *Q. bicolor*, *Q. stellata*, *Q. macrocarpa*, *Q. lyrrata*, and *Quercus durandii* [18,26].

In Europe, the largest forests producing high-quality oaks are found in France. Forests cover 27% of the total area of France, and approximately 9% of these are oak forests. The regions of Le Fôret du Centre, Nevers, Tronçais, Allier, and Limousin in the Massif Central and Vosges in the northeast of the country are particularly important producers. Although French oak is the most highly valued in Europe, other producer regions include Hungary, Poland, Russia, Italy, and, in the Iberian Peninsula, the Basque Country. In fact, until the 1930s, the oak that was most widely used in the châteaux of Bordeaux came from Russia rather than France. However, *Q. petraea* and *Q. robur* are associated with France and not the rest of Europe.

In general, American oaks differ from European species because they have higher density and resistance and lower porosity and permeability than European species [25]. In addition, American woods have larger tylosis, which allows this wood to be cut by sawing without compromising the watertightness, which leads to a better use of wood. Heartwood of *Quercus* is composed of macromolecules that are polymers of cellulose, hemicellulose, and lignin, representing 90% of dry wood. In addition, there is an extractable fraction that are soluble compounds released to wine during aging; this part represents approximately 10% of dry wood and is variable depending on species. The extractable fraction (ellagitannins, low molecular weight compounds and volatile compounds) in wood depends not only on variety but also on many other factors, such as sylvocultural factors, geographic origin, and individual tree and cooperage processing, with high variability in content (Tables 1 and 2). In general, *Q. robur* has the highest content of ellagitannins followed by *Q. petraea* and finally *Q. alba*. In addition, *Q. alba* also tends to have lower content of low molecular weight compounds (Table 1) and *Q. robur* lower aromatic compounds (Table 2).

Table 1. The range of ellagitannins and low molecular weight phenolic (LMWP) compounds found in green, seasoned, and toasted wood. * sum of castalagin, vescalagin, and A, B, C, D, and E roburins; ** sum of acids (ellagic, gallic, syringic, vanillic, and ferulic), aldehydes (coniferaldehyde, sinapaldehyde, syringaldehyde, and vanillin) and cumarins (scopoletin and aesculetin); *** sum of gallic acid and elagic acids.

Treatment	Species	Concentration Range Ellagitannins (mg/g) *	References	Concentration Range LMWP (μg/g) *	References
Untreated (green wood)	Quercus pyrenaica Willd.	28.12–32.72	[27–29]	265–1061	[23,27,28,30]
	Quercus faginea Lam.	32.51	[28,29]	407	[28,30]
	Quercus frainetto Ten.	-	-	-	-
	Quercus oocarpa Liebm.	-	-	-	-
	Quercus humboldtii Bonpl.	1.94	[16]	365	[16]
	Castanea sativa Mill.	-	-	-	-
	Robinia pseudoacacia L.	-	-	-	-
	Prunus avium L.	-	-	-	-
	Prunus cereaus L.	-	-	-	-
	Fraxinus americana L	-	-	-	-
	Fraxinus excelsior L.	-	-	-	-
	Quercus petraea	8.65–32.10	[16,28,29]	225–752	[16,28,30]
	Quercus robur	28.41–44.01	[28,29]	310–647	[28,30]
	Quercus alba	3.48–5.96	[16,29]	237–486	[5,16]
Seasoned	Quercus pyrenaica Willd.	2.81–77.9	[8,31–33]	475–4304	[7,28,31,33–35]
	Quercus faginea Lam.	24.11–26.97	[8,28]	760–1422	[7,28]
	Quercus frainetto Ten.	108	[36]	3800 ***	[36]
	Quercus oocarpa Liebm.	33.9	[36]	5500 ***	[36]
	Quercus humboldtii Bonpl.	1.61	[17]	832	[17]
	Castanea sativa Mill.	4.74–76.3	[31,33,36–38]	1155–14,430	[31–34,36—41]
	Robinia pseudoacacia L.	nd	[42,43]	41–408	[40,42]
	Prunus avium L.	nd–0.04	[31,43,44]	6–620	[31,44]
	Prunus cereaus L.	-	-	228	[40]
	Fraxinus americana L.	nd	[43,45]	98	[45]
	Fraxinus excelsior L.	nd	[43,45]	53	[45]
	Quercus petraea	1.98–80.62	[8,17,28,31,32,36,38]	368–3400	[7,17,28,31,34,38]
	Quercus robur	3.93–87.4	[8,31,32,36]	647–4166	[7,28,31,34,45]
	Quercus alba	0.88–35.64	[8,17,31,32,36]	469–1064	[7,17,31]

Table 1. *Cont.*

Treatment	Species	Concentration Range Ellagitannins (mg/g) *	References	Concentration Range LMWP (µg/g) *	References
	Quercus pyrenaica Willd.	4.32–47.05	[8,32,33,46]	607–20,500	[7,33,35]
	Quercus faginea Lam.	9.34	[8]	2132	[7]
	Quercus frainetto Ten.	-	-	-	-
	Quercus oocarpa Liebm.	-	-	-	-
	Quercus humboldtti Bonpl.	0.12	[17]	2464	[17]
	Castanea sativa Mill.	0.66–10.51	[33,37]	1353–35,282	[7,33,35,37,40,47,48]
Toasted	*Robinia pseudoacacia* L.	nd	[42,49]	6–2496	[40,42,44,46]
	Prunus avium L.	nd	[43,46]	90–3378	[44,46,48]
	Prunus cereaus L.	-	-	445–1578	[40]
	Fraxinus americana L.	nd	[43,45]	1915–3062	[45]
	Fraxinus excelsior L.	nd	[43,45]	1922–3585	[45]
	Quercus petraea	3.53–56.76	[8,17,32,46]	856–4420	[7,17,32,46]
	Quercus robur	7.72–11	[8]	2067–8225	[7,40,47,48]
	Quercus alba	nd–5.89	[8,17,32,46]	460–3620	[7,17,32,46,47]

nd: not detected.

Table 2. The range of volatile compounds expressed as µg/g wood found in green, seasoned, and toasted wood.

Treatment	Species	Guaiacol	Eugenol	Furfural	Trans-β-Methyl-γ-Octalactone	Cis-β-Methyl-γ-Octalactone	Vanillin	References
Untreated (green wood)	Q. pyrenaica Willd.	0.16–0.24	1.47–5.71	1.35–2.56	0.84–29.37	14.35–59	3.42	[27,50]
	Q. petraea	0.09	2.05	1.19	18.4	36.9	2.19	[50]
Seasoned	Q. pyrenaica Willd.	nd–1.25	nd–7.28	1.94–19.7	nd–33.8	5.3–68.2	1.6–25.24	[27,51–55]
	Q. faginea Lam.	0.29	1.98	17.6	1.74	15.5	10.6	[51]
	Q. humboldtti Bonpl.	0.1	2.65	nd	nd	0.02	1.97	[17]
	Castanea sativa Mill.	nd–0.38	0.71–4.47	2.27–6.72	nd–0.23	nd–0.34	2.90–24.40	[53,55–57]
	Robinia pseudoacacia L.	nd–0.86	nd–0.21	0.45–0.92	nd	nd	1.65–3.48	[55–57]
	Prunus L.	nd–0.53	nd–0.11	0.49–0.66	nd	nd	0.13–2.42	[55–57]
	Fraxinus americana L.	0.08–0.13	0.19–0.57	0.54–0.8	nd	nd	7.25–10.3	[55,57]
	Fraxinus excelsior L.	0.11–0.22	0.44–0.94	1.21–1.31	nd	nd	1.39–14.7	[55,57]
	Q. petraea	nd–1.3	0.57–6.5	3.4–19.9	0.09–14.7	0.42–55.9	2.0–18.8	[50,51,53–55]
	Q. robur	0.08–0.11	1.01–1.58	4.51–10.8	2.87–3.98	2.83–22.9	1.21–15.9	[51,53]
	Q. alba	0.04–3.3	1.38–5.9	1.2–4.68	2.52–5.0	22.3–32.5	6.8–7.9	[51,54,55]
Toasted	Q. pyrenaica Willd.	0.11–8.91	nd–14.6	19.6–4082	nd–58.6	0.10–212	8.69–235	[51,54,55,58–60]
	Q. faginea Lam.	0.36	2.35	96	1.08	3.25	258	[51]
	Q. humboldtti Bonpl.	3.74	3.34	533.53	0.06	0.3	22.36	[17]
	Castanea sativa Mill.	0.46–5.30	2.13–3.23	431–1675	nd	nd	7.15–143	[55,57]
	Robinia pseudoacacia L.	0.52–6.05	0.40–2.36	20.7–840	nd	nd	19.2–106	[55,57]
	Prunus L.	0.91–1.71	0.74–1.50	23–175	nd	nd	45–91.7	[55,57]
	Fraxinus americana L.	5.97–11.9	1.58–3.00	26.5–63.6	nd	nd	76.1–160	[55,57]
	Fraxinus excelsior L.	6.47–14.07	1.59–3.21	26.5–82	nd	nd	76.3–187	[55,57]
	Q. petraea	0.17–3.4	0.83–4.0	10.3–963	0.01–14.6	0.05–22.8	3.0–370	[51,54,55,60]
	Q. robur	0.17–0.53	1.01–1.37	8.90–10.8	3.41–3.98	2.83–22.9	130–172	[51]
	Q. alba	1.22–7.3	1.29–11.6	4.04–1539	3.29–7.4	16.1–45.5	7.5–102	[51,54,55,60]

Furthermore, the use of oak barrels in the production of quality wines implies long periods and a high economic cost for wineries. For this reason, alternative techniques to ageing in oak barrels have been used for over 15 years and these were developed to give wood characteristics to the wine in a faster, cheaper, and simpler way. They are based on the addition to the wine of pieces of wood of very different sizes and shapes (splinters, cubes, staves). These alternative products have been widely used for a long time in the producing countries of the New World, but their use has spread, above all, since the main wine producer, Europe, changed its legislation to admit their use. This maturation practice was approved by the International Oenological Codex of the International Organization of Vine and Wine (OIV) (OENO 9/2001) and by the Official Journal of the European Union (CE 1507/2006).

2. Oak Species Not Traditionally Used in Cooperage

The use and/or study of alternative oaks (other species of the genus *Quercus*) is proposed as a solution to the search for new sources of quality wood for cooperage that provide wines with differentiated notes appreciated by the consumer. For this reason, a market opportunity has arisen for oak species not traditionally used in cooperage, such as *Quercus pyrenaica*, *Quercus faginea*, *Quercus frainetto*, *Quercus oocarpa*, and *Quercus humboldtii* and others that are less well-known, such as *Quercus serrata*, *Quercus mongolica* or *Quercus denta* (Figure 1).

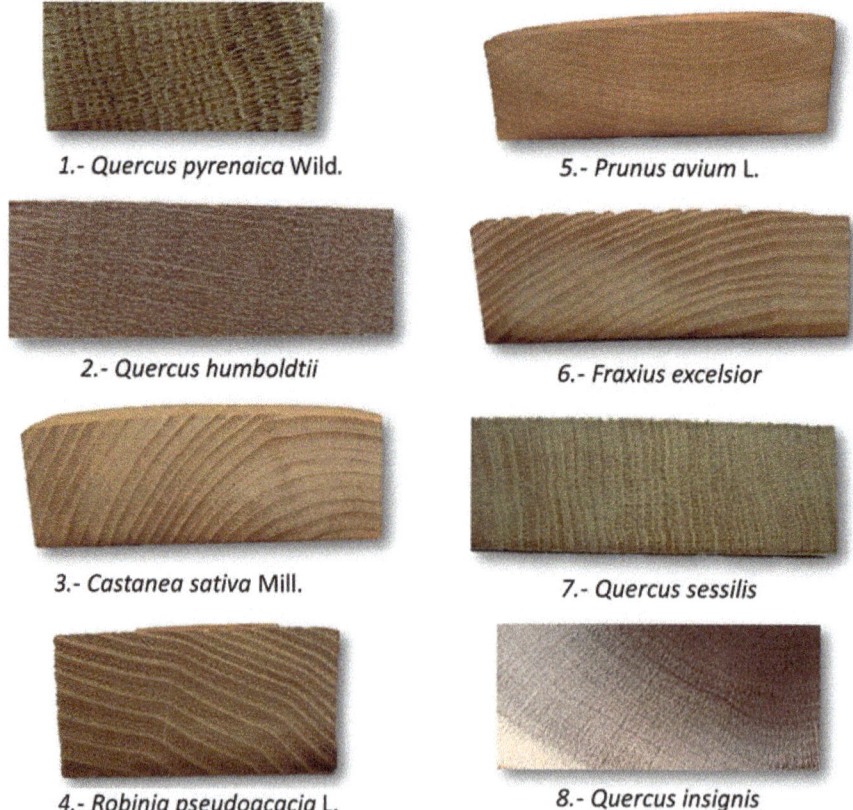

1.- Quercus pyrenaica Wild.

5.- Prunus avium L.

2.- Quercus humboldtii

6.- Fraxius excelsior

3.- Castanea sativa Mill.

7.- Quercus sessilis

4.- Robinia pseudoacacia L.

8.- Quercus insignis

Figure 1. Cross-sections of some of the non-standard species in cooperage in comparison with *Q. sessilis*.

2.1. Pyrenaica Oak (Quercus Pyrenaica Willd)

It is distributed throughout the western Atlantic–Mediterranean regions (West France, Portugal, Spain, and North Morocco) through a wide range of altitudes, from sea level to over 2000 m. This wood is known as "rebollo" or "melojo", and is mostly located in Spain (Allué, 1995) with a forest mass of 1,090,716 ha, the majority of which is found in the region of Castilla y León [61] (Figure 2). Traditionally, this wood has been used in Spain for railway sleepers and ships and, in recent years, especially as firewood from low forest cover, an arboreal mass composed of feet coming from buds or roots. This has resulted in a progressive degradation of the characteristics of some of these forest areas, such as a high percentage of trees with a diameter of <40 cm and knotty, twisted, or short-boled trees. Therefore, their use for manufacturing barrels is very limited due to the high number of poor quality trees for cooperage. However, its structural properties (mesh, grain, density, and permeability) are also appropriate for oenological use [27].

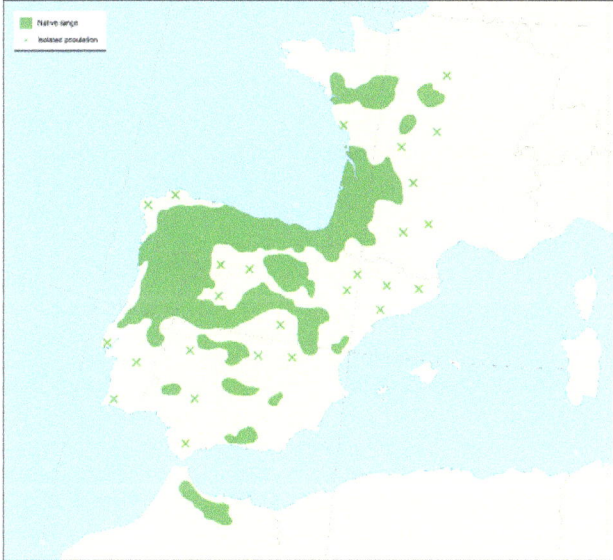

Figure 2. *Quercus pyrenaica* Willd distribution map [62]. https://commons.wikimedia.org/wiki/File: Quercus_pyrenaica_range.svg.

Various studies from 1996 to the present place value on this wood's content of ellagitannins, low weight, and aromas [7,8,27–32,34,35,46,50–53,55,58,59,63–65], as well as its use for containers of alcoholic drinks, such as brandy and other spirits [66–70] and wines [9,10,58,60,71–79]. However, the supply of quality wood for the manufacture of barrels is insufficient, so this wood can be used for the manufacture of alternative products in the short and medium term. With proper management, these forests could supply wood for the manufacture of barrels in the future. Consequently, most studies on the behaviour of this wood during wine ageing have been carried out with alternative products (in particular chips or staves) [58,60,71–77,79], observing that wines aged with this wood present good final characteristics. The resulting wines are closer to those aged with French oak than those aged with American oak, thus meaning this wood is suitable for producing quality wines [9,10,74]. In addition, peculiarities have been reported, such as that wines aged with barrels of this oak species had high levels of eugenol, guaiacol, and other volatile phenols, while the contents of cis-β-methyl-γ-octalactone or maltol are similar to those of wines aged with *Q. alba* [76]. However, Fernandez de Simón et al. [60] observed that the Tinta del País variety, in addition to having a higher concentration of eugenol, also had a higher content of cis-β-methyl-γ-octalactone than the same wine treated with French and

American oak, especially when staves were used. In tasting, wines aged in *Q. pyrenaica* wood barrels are more appreciated than the same ones aged in American or French oak barrels [76]. Gallego et al. 2012 [79] also observed that the wines aged with chips and staves from *Q. pyrenaica* oak were better considered than American or French ones, showing higher aromatic intensity and complexity, and woody, balsamic, and cocoa notes. Connick et al. [77] carried out a sensorial analysis and found that wine aged in contact with *Q. pyrenaica* chips only differed to wine aged with *Q. petraea* for woody character and attained a higher score. However, a wine from two Portuguese red grape varieties (Tinta Roriz, 80% and Touriga Nacional, 20%), aged with chips of *Q. petraea* versus *Q. pyrenaica*, reported that, from a sensory point of view, the wine with French oak chips showed a tendency for higher aroma scores than those aged in contact with *Q. pyrenaica* oak [71]. As regards oxygen, a very important factor in the ageing of wines, Del Álamo et al. [73] reported that the same red Tempranillo variety aged with *Q. pyrenaica* required less oxygen than the same one aged with traditional oak species (*Q. petraea* and *Q. alba*). Similarly, Gonalves and Jordão 2009 [75] recorded that Syrah wines aged in contact with *Q. pyrenaica* oak species had higher antioxidant activity values than those aged in contact with the American oak species.

2.2. Quercus Faginea Lam

A wood studied from 1996 [30] to the present [80] for oenological purposes. This species is endemic to the Iberian Peninsula and North Africa. Its common name is Quejigo, with about 269,000 ha distributed mainly in Castilla-La Mancha, Castilla y León, Aragón, and Cataluña on the peninsula [61,81] (the protected surface area is under 4%, due to its dispersion and abundance in Spain [81]). It is a medium-sized deciduous or semi-evergreen tree growing to a height of 20 m and a diameter of 80 cm. This species once covered (during the 15th and 16th centuries) much of the Iberian Peninsula, and the wood was valued and intensively exploited for naval construction [82]. It was traditionally used for the production of charcoal, and it has also been used in the manufacture of beams for construction due to its strength and resistance. *Q. faginea* wood has a white yellowish sapwood and brown yellowish heartwood, high density, and considerable mechanical strength [80]. Its composition in ellagitannins, low molecular weight compounds, and volatile compounds in wood has been studied [7,8,10,28–30,51,63,80], as well as its interaction with wine [9,10,78]. The antioxidant activity is very high, with an IC50 of 3.3 µg/mL for heartwood, as compared to standard antioxidants (an IC50 of 3.8 µg/mL for Trolox) [80]. Miranda et al. [80] reported that *Q. faginea* is a very good candidate for cooperage due to it being a source of compounds with antioxidant properties. However, after studying the volatile composition of different species of Spanish oak, both in traditional species (*Q. robur* and *Q. petraea*) and new species (*Q. faginea* and *Q. pyrenaica*), Cadahía et al. [51] propose that, while *Q. pyrenaica* may be considered suitable for wine ageing, *Q. faginea* is not. In addition, a wine was in contact with this oak for 21 months and then compared with other species, especially traditional ones, and it was observed that wine aged with *Q. faginea* was the least preferred by the tasting panel and always the one that obtained the lowest scores in almost all descriptors [10]. Fernández de Simón et al. 2003 [9] studied the low molecular weight phenolic compounds in red Rioja wine aged during 21 months in barrels made of *Q. Faginea* oak and other woods (*Q. pyrenaica*, *Q. robur*, *Q. petraea*, and *Q. alba*), showing by means of a discriminant analysis that wines aged with *Q. faginea* could be discriminated from the rest by function 2, which was related to trans-resveratrol, p-hydroxybenzaldehyde, syringic acid, ellagic acid, and 5–HMF.

2.3. Quercus Frainetto Ten

This species is native to the Balkan Peninsula and also present in South Italy and Northwest Turkey (Figure 3). Despite also being known as Hungarian oak, its presence in Hungary is sporadic [83]. In Greece, it is a vital timber tree and frequently managed as coppice forest for both firewood and timber in combination with grazing. In the other countries in which it grows, it is most often used for firewood, although the quality of the wood is similar to *Q. petraea*. Because of the rather high durability of its wood, *Q. frainetto* has sometimes been used as construction material in civil engineering and mining. Vivas [21] is studying this species with the aim of using it in cooperage. This wood has an ultra-structure that is comparable to French oaks, and its lindens are similar to those of *Q. alba*. However, in the manufacture of barrels with *Q. frainetto*, the staves have been found to need longer heating during taming, which could be due to their high density [21]. This species also has a high content in ellagitannins [36]. As regards the gustatory quality of the wood extracts of this species, it has high bitterness and particular and indefinable aromas, but both attributes can be cushioned by the natural drying and toasting of the wood [21].

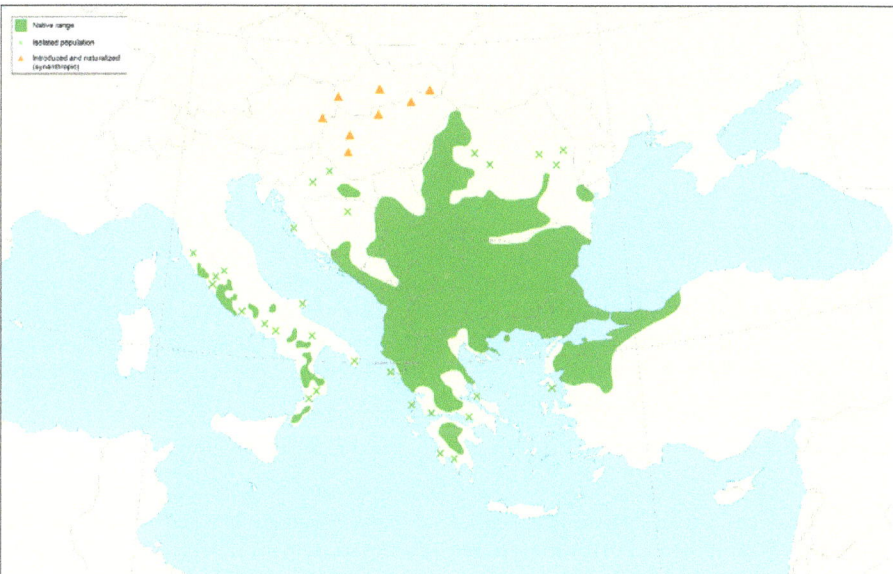

Figure 3. *Quercus frainetto* Ten. distribution map [62]. https://commons.wikimedia.org/wiki/File: Quercus_frainetto_range.svg.

2.4. Quercus Oocarpa Liebm

This species is also used in wine ageing [36] and extends naturally from Veracruz, Mexico, through Chiriquí, Panama, Guatemala, and Costa Rica, where it is found in Monteverde, Puntarenas; Cordilleras de Tilarán and Central; Escazú, San José; Muñeco, Cartago; and the Cordillera de Talamanca [84]. Vivas [21] proposes this as a new species when observing that it presents an ultra-structure that is comparable to French oaks with a clear succession of early and late wood, forming an annual growth and, with respect to its lime trees, observed that *Q. oocarpa* was comparable to *Q. alba*. This species presents only monomers of ellagitannins, since during its analysis no dimer was found [36]. Regarding the gustatory quality of the extracts of these woods, the quality of the *Q. oocarp* was similar to that of *Q. petraea* [21].

2.5. Quercus Humboldtii Bonpl

This is one of the main forest species in the woods of Colombia [85]. This white oak is a neotropical species found in the Three Mountains range, from 750 m to 3450 m above sea level, in 18 departments of the Colombian Andes (Antioquia, Bolívar, Boyacá, Caldas, Caquetá, Cauca, Chocó, Cundinamarca, Huila, Quindío, Risaralda, Nariño, North of Santander, Santander, Tolima, Valle del Cauca, Cesar, and Córdoba) [85]. The hardwood is hard, heavy, and easy to work and its density is 0.9–1 g/cm^3. Traditionally, it has been used for making posts, railroad ties, handles for tools, wooden rollers, charcoal, and firewood [86]. In addition, this species is normally used in barrel-making; specifically, it has been utilised by two companies since the middle of the 20th century. Cooperage products made from this "White oak" have normally been used to age alcoholic beverages, such as rum [87] or brandy. Recently, three studies on the composition of this oak when green and before and after toasting have been published [16,17,88]. The phenolic composition (ellagitannins and low molecular weight phenols) of green *Q. humboldtii* was characterized and compared to traditional oak wood species, with the most abundant phenolic acids, aldehydes, and ellagitannins being the same as in *Q. alba* and *Q. petraea*, and with a phenolic composition closer to that of the American ones [16]. The study on syringaldehyde and vanillin contents showed a similar vanillin concentration to *Q. Faginea* in toasted wood and a balanced syringaldehyde/vanillin relationship, a marker usually used to characterize oak wood quality [88], when seasoned and toasted. *Q. humbolditti* had comparable low molecular weight phenols to woods of *Q. petraea* and *Q. alba*. Its ellagitannin composition was similar to that in *Q. alba*, and its volatile composition differed from that of *Q. petraea* and *Q. alba*, since it had the highest concentration of 5-methyl furfural, furfuryl alcohol, guaiacol, 4-ethylguaiacol, 4-vinylguaiacol, cis and trans-isoeugenol, and syringol and the lowest furfural, 5-hydroxymethylfurfural, and *cis*-β-methyl-γ-octalactone concentrations [17]. When this wood is used as an alternative for ageing wines compared to traditional species, the wines macerated with *Q. humboldtii* chips showed higher concentrations of 5-methylfurfural, guaiacol, isoeugenol, trans-isoeugenol, and syringol and lower furfural, 5–HMF, *trans*-β-methyl-γ-octalactone, and *cis*-β-methyl-γ-octalactone content [89]. In the sensorial analysis, there were no negative comments from the tasters about the wine macerated with Colombian oak; in addition, few significant differences in the sensorial analysis were observed in these wines compared to those aged with traditional oaks. Therefore, *Q. humboldtii* oak has an interesting oenological potential as an alternative species for coopering [89].

3. Woods Not Traditionally Used in Cooperage Different to Oak

The growing demand for wood for cooperage and the search for new opportunities to give wines and their derivatives a special personality have led to the use of woods other than oak, some of which have been used for many years. This is how wood from species such as *Castanea sativa* Mill. (chestnut), *Robinia pseudoacacia* L. (false acacia), *Prunus avium* L. and *Prunus cereasus* L. (cherry), *Fraxinus excelsior* L. and *F. americana* L. (*European ash* and *F. americana* L.). (European and American ash, respectively), and *Morus alba* L. and *Morus nigra* L. (Mulberry) have been proposed as alternatives to oak (Figure 1). In addition to those mentioned above, experiments in wines have been made with other types of wood, such as *Juglans regia, Juniperus communis, Pinus heldreichii var. Leucodermis, Prunus armeniaca, Fagus Syvatica*, and *Alnus glutinosa* [48,90], but, to date, few trials have been carried out. Moreover, many producers prefer using local woods in order to reduce costs [43], and, recently, some wine cellars have ordered barrels with some non-oak staves included from cooperages.

3.1. Castanea Sativa Mill

This species of the Fagaceae family can be found in southern Europe and Asia (China) (Figure 4). Chestnut is widely cultivated for its tasty edible fruits. Its starch is used in industrial applications, such as paper, plastics, textiles, food, pharmaceuticals, and cosmetics, and its wood is of interest for the manufacture of stakes. Moreover, this species has been widely used for oenological purposes in the Mediterranean area in the past due to its widespread availability and low cost [91]. As mentioned above, it is the only species alongside *Quercus* that has been accepted for use by the OIV. It seems that there is a growing interest in the use of this wood in the ageing of different drinks, which is why numerous studies have been carried out on the characteristics of this wood [31,33,34,36–41,43,47,51, 53,54,92,93] and its use for the purpose of ageing spirits [36,67–70,91,94–103], vinegars [53,104,105], and wines [57,106–113]. Chestnut wood barrels prove to be suitable for the ageing of wine liqueurs, as they improve the chemical composition and the sensory properties of the alcohol of the aged wine, showing higher content of total phenolics and of low molecular weight compounds and higher antioxidant activities [100]. The sensory properties found in spirits aged in chestnut wood demonstrate the potential of this wood for the ageing of alcoholic beverages [69], with the heat treatment in cooperage having a very significant influence on the majority of low molecular weight extractable compounds by brandies aged two years in chestnut barrels [39]. Chestnut wood proved to be a suitable alternative to Limousin oak for ageing brandies, showing a high quality, with a faster evolution of brandies and more economical ageing as the price is lower [91]. However, chestnut does not seem to be the most suitable for vinegar ageing, as the best results are found when oak or cherry are used [104,105]. Regarding the ageing of wine in barrels made of this wood, it has been observed that they are suitable for short periods, but not for prolonged ageing, due to the high porosity [108–113]. Thus, Rosso et al. [113] observed a low content of oxidizable polyphenols in wines aged with chestnut, indicating that this type of wood causes a more oxidative environment than oak and is, therefore, less suitable for prolonged ageing. Alañon et al. [108] reported that wines aged for long periods in chestnut present off-flavours (4-ethylphenol and 4-ethylguaiacol) and oxidation problems. However, chestnut is an excellent flavouring wood for short periods of ageing in barrels, as very balanced wines are obtained [108]. Arfelli et al. [106] observed that red Sangiovese wine aged in old chestnut barrels were more fruited and tannic than in Allier, while the latter were less astringent, more balanced, and had more vanilla notes.

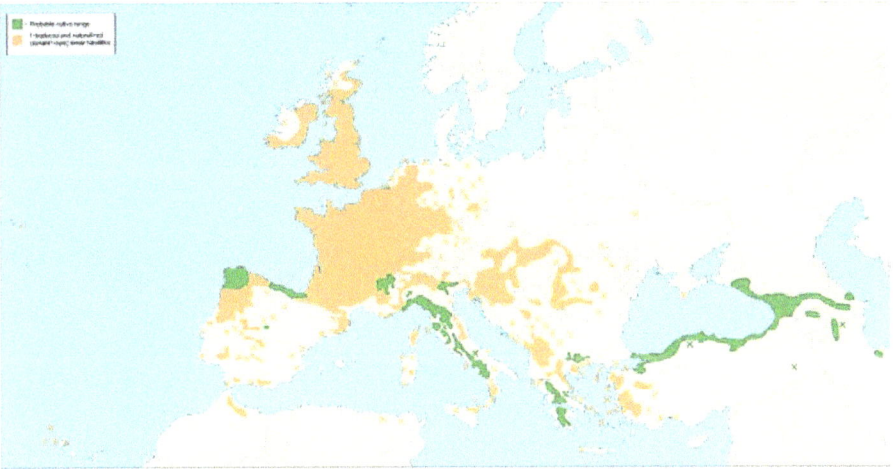

Figure 4. *Castanea sativa* Mill. distribution map [62]. https://commons.wikimedia.org/wiki/File:
Castanea_sativa_range.svg.

3.2. Robinia Pseudoacacia L., (False Acacia)

This species, originating in the eastern United States and introduced into Europe, is often referred to as acacia, but its proper name is robinia [42] (Figure 5). It is considered a rapid-growth species with adaptive plasticity compared to others [114]. Traditionally, it has been used for the production of poles and pulp, as well as for other uses, such as erosion control and fodder. It has now been proposed for cooperage purposes, as robinia barrels are approximately 10% cheaper than French oak though still more expensive than American oak. Acacia wood is hard, with a low porosity [115]. In the last 10 years, research papers have been published focusing on the characterization of this wood for cooperage purposes and its use with alcoholic beverages, especially wines [71,110–113,116–119] and vinegars [104,105,120]. Red wines aged in acacia barrels have higher notes of smoky, spicy, and fruity, and may be related to their richness in mono and dimethoxyphenols, acetosyringone, and ethyl vanillate [111,121]. Fernández de Simón et al. [111] observed that, after the use of barrels of the species cherry, chestnut, acacia, ash, and oak, the wines with the highest scores were those aged with acacia and oak [111]. Acacia barrels also had a positive influence on the quality of Istria wines, as they were the best-rated with the highest amount of simple volatile phenol compounds [116]. In the case of white wines, the preference of this wood over others, such as cherry and even over American and French oak, was also observed [119]. The use of acacia for vinegar ageing is increasing due to the air transfer efficiency that favours a good rate of acetification [122], since, of all of the woods studied, the acacia barrels were observed to be those with the highest oxygen permeability. However, regarding the aromatic notes of the vinegars aged with wood, Callejon et al. [105] suggest that the best woods are cherry and oak, not acacia.

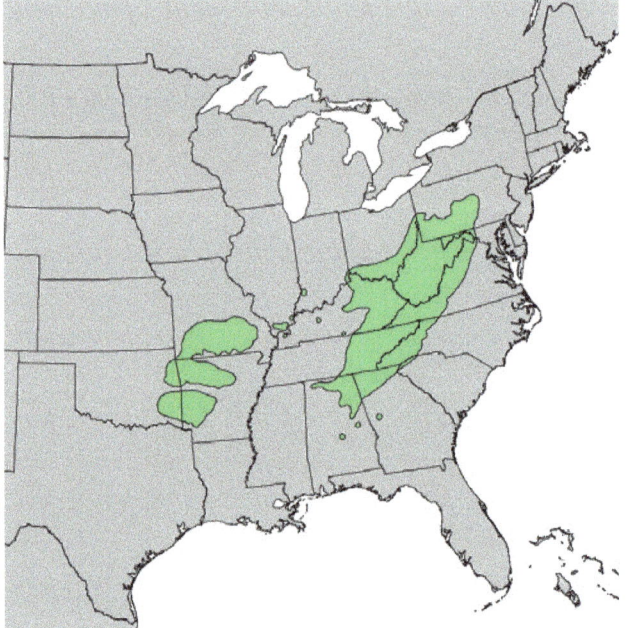

Figure 5. *Robinia pseudoacacia* L. distribution map [123]. https://commons.wikimedia.org/w/index. php?curid=29169867.

3.3. Prunus

The cherry species studied for this purpose are *Prunus avium* L. and *Prunus cerasus* L., related to each other and native to Europe and western Asia (Figure 6). The cherry has been extensively studied in recent years in order to know its characteristics [31,43,44,55–57,92,124] and the effect it provides during the ageing of different drinks, such as wine, distillates, and vinegars [40,48,71,105,110–113, 117,119,125,126]. This wood has a high porosity and oxygen permeation, and is usually used for short ageing times [40]. Fernández de Simón et al. [110], after studying white, rosé, and red wines aged using barrels and chips, observed that 6 (aromadendrin, naringenin, taxifolin, isosakuranetin, eriodictyol, and prunin) of the 68 identified nonanthocyanic phenolic compounds were only identified in wines aged with cherry wood. Thus, the nonanthocyanic phenolic profile could be a useful tool to identify wines aged in contact with this wood. In addition, significant differences were found in certain compounds with respect to *Q. petraea* and *Q. alba*. Delia et al. [119] showed that the white wine aged in contact with cherry chips showed similar overall appreciation scores to those obtained for the wines aged with *Q. alba* and *Q. petraea* chips. De Rosso et al. [113] also found some special characteristics in wines aged in untoasted cherry barrels compared to other woods (acacia, chestnut, mulberry, and oak), suggesting that this wood allows for greater oxygen penetration through its staves. However,

Torrija et al. [122] observed that the most permeable to oxygen was acacia. Cerezo et al. [104] observed that vinegars from red wines after their acetification with better scores in the sensory analysis were those from ageing with this wood together with oak, presenting better notes of global impression and red fruits. De Rosso et al. [113] also found some special characteristics in wines aged in a cherry-barrel compared to other woods (acacia, chestnut, mulberry, and oak), suggesting greater oxygen penetration through their staves; and, therefore, proposed their use for shorter ageing times.

Figure 6. *Prunus avium* L. distribution map [62]. https://commons.wikimedia.org/wiki/File:Prunus_avium_range.svg.

3.4. Fraxinus

This genus of the family of oleaceae, generally known as ash, is found in the geographical area of the *Fraxinus excelsior* L. extending throughout Europe, Asia Minor, and North Africa, preferably in oceanic climates (Figure 7). It reaches heights of up to 40 m, and specimens from 20 to 30 m are common. This species only grows properly in areas where the climate and soil conditions provide a good water supply throughout the year. Ash is highly appreciated. The highest quality logs are destined for the veneer industry, where they reach their maximum price. Ash is also highly valued in the sawmill and cabinet-making industries. *Fraxinus americana*, L. a native of North America, is found mainly in the eastern United States and has been introduced in Cuba and Romania [127,128]. It is a large tree approximately 36 m high and 182 centimetres in diameter, and is highly appreciated thanks to the qualities of its wood, which is moderately heavy, strong, rigid, hard, and resistant to shocks. Because of these characteristics, it is mainly used for handles, ropes, oars, vehicle parts, baseball bats, and other sporting goods, as well as for veneers, sawn wood, and canoes. Heartwood from *Fraxinus*, both excelsior and American, has been considered as a possible source of wood for ageing wines [57,110,111], so its composition has been studied from the oenological point of view [43,45,48,124]. Wines aged in ash barrels differed from the rest due to their high content of 3-ethyl and 3,5-dimethylcyclotene, o-cresol, α-methylcrotonalactone, and vanillin and their low content of furanic derivatives, the latter like wines aged in cherry [111]. In spite of showing greater quantities of vanillin, after a sensorial analysis, the wine aged with oak had the highest scoring vanilla notes [111]. The polyphenolic profile of wines aged in contact with ash has not shown any specific polyphenols provided by this wood, with no unusual compounds being found when the wine was aged with oak [110], although ash has shown to have compounds not present in oak [45].

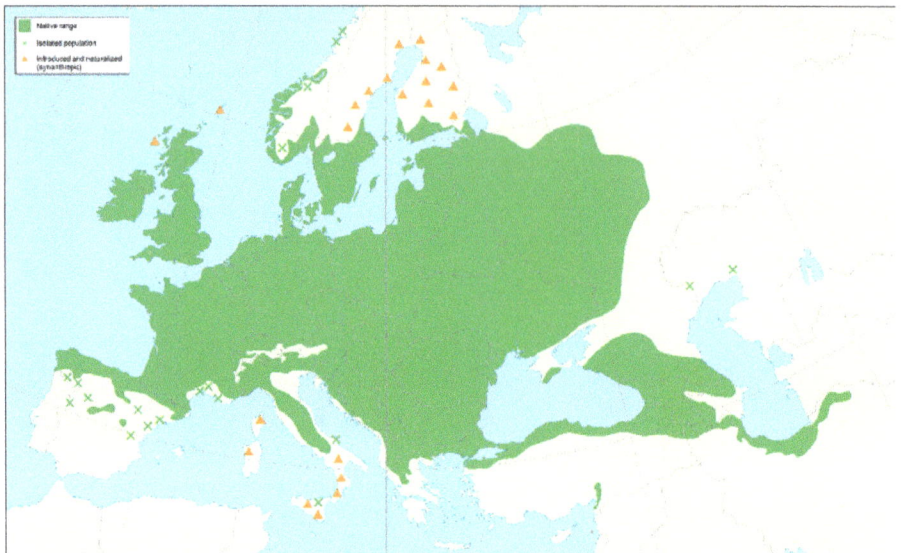

Figure 7. *Fraxinus excelsior* distribution map [62]. https://commons.wikimedia.org/wiki/File: Fraxinus_excelsior_range.svg.

3.5. Morus (Mulberry)

The mulberry species that have been considered as possible new woods in wine ageing have been *Morus alba* L. (known as white mulberry) and *Morus nigra* L. (known as black mulberry). *Morus alba* L. is a native of China but widely planted and naturalized in many warm temperate regions. *Morus nigra* L. is a native of western Asia but mostly cultivated in Europe and Asia [129]. In general, *Morus* L. (Moraceae) is found in Asia, Africa, Europe, and North, Central, and South America [130] and grows in various forest types from sea level up to 2500 m [131]. Morus species are economically important to the silk industry, as they are host plants for the silkworm (*Bombyx mori* L.) larvae [132]. Additionally, species have been cultivated in many parts of the world for their edible fruits and as ornamental trees. Moreover, *M. alba* is the main species for making traditional bowl-shaped musical instruments [133]. Karami et al. [133] studied the anatomical differences and similarities between these two species of wood, showing that small differences exist between them in vessel distribution and frequency and the existence of aliform axial parenchyma cells. The main differences reported by these authors were a semi-ring porous distribution of vessels in *M. alba*, and fewer vessels and a lower presence of aliform parenchyma in *M. nigra*. The wood of these species is tender and elastic, with medium porosity, and is characterized by the low release of compounds [134]. These species have been less-studied from the point of view of their characterization and use for the ageing of beverages than those previously mentioned. Rosso et al. [92] studied extracts (50% water/ethanol *v/v*) with 60 g/L of different woods (acacia, chestnut, cherry, mulberry, and oak) and observed that the lowest contents of volatile compounds were found in mulberry, with little eugenol and no methoxyeugenol though high (negative) fatty acids. Flamini et al. [56] extracted the compounds from the same woods and in the same dosage as in the previous study, not only with a 50% water/ethanol solution but also with model wine (12% ethanol with tartrate buffer pH 3.2), showing that the mulberry wood extract had a low presence of volatile benzene compounds and is probably more suitable for ageing wines.

Gortzi et al. [90] studied two Greek red wines (Syrah and Cabernet) aged with white Mulberry wood chips. They observed that the total polyphenol content (mg/L) in Syrah wines aged with *M. alba* was lower than those aged with *Q. alba* when the dosage of alternatives was 1 g and very similar when it was 2 g. The opposite occurred when the grape variety used was Cabernet. Gortzi et al. [90] saw that the concentration of resveratrol and catechin in all of the wines studied was much higher when aged with *M. alba* chips than when aged with *Q. alba*. The sensory test showed that, after 20 days' ageing with *M. alba*, Syrah wines presented better scores than the same ones aged with *Q. alba*; however, the scores of the Cabernet wines after ageing with these two woods were similar [90]. Mulberry extract (*M. nigra* heartwood) (2 g/L of wood in 40% water-ethanol) had higher a polyphenol content and antioxidant activity than the extracts from the *Quercus robur*, *Robinia pseudoacacia* L., and *Cotinus coggygria Scop* wood species [116]. Rosso et al. [113] observed that a red wine (Raboso Piave var.) aged during 9 months in blackberry (*M. alba*) 225-L barrels presented a significant decrease in fruity-note ethyl esters and ethylguaiacol and the high cession of ethylphenol (a horsey-odour defect). Therefore, these authors concluded that this wood is hardly suitable for wine ageing.

4. Chemical Composition of the Extractable Fraction of the Different Woods

The chemical composition, especially the extractable fraction, of oak wood can decisively condition its oenological quality, as it contributes to characteristics such as the colour, smell, flavour, and body of the final wine. Ellagitannins, low molecular weight compounds, and volatile compounds are the main constituents of this fraction in oak. Table 1 shows the total concentration of ellagitannins and the total content of low molecular weight phenolic compounds in green, dried, and toasted wood of *Quercus* specimens but of different species than the traditional ones (*Quercus*: *pyrenaica*, *faginea*, *frainetto*, *oocarpa*, and *humboldtti*) and woods other than oak (*castanea sativa*, *robina pseudoacacia*, *prunus avium*, *prunus cereaus fraxinus americana*, and *fraxinus excelsior*). In general, the ranges found for total ellagitannins and low molecular weight phenolic compounds in the same species are broad due to factors such as the type of treatment and the intensity and variability within the species, and are shown in Table 1.

As can be seen in Table 1, the species *Robinia pseudoacacia* (acacia), *fraxinus americana* (ash), and *fraxinus excelsior* (ash) do not contain ellagitannins in their composition. With regard to cherry, no studies have been found on the species *Prunus cereaus*. The ellagitannins in *Prunus avium* were not detected in the studies by Sanz et al. [43,44], with only very small quantities of castalagina (0.04 mg/g) and vescalagina (4.19 µg/g) being detected in the work of Alañon et al. [31], insignificant quantities with respect to the habitual contents that exist in the woods used in cooperage. As regards woods other than oak, the eight ellagitannins have only been found in the species *Castanea sativa* Mill (chestnut). The ranges of ellagitannins in dried chestnut are between 4.74 and 76.3 mg/g and in toasted vary from 0.66 to 10.51 mg/g (Table 1). As with oak [17,32,135], toasting decreases the concentration of these ellagitannins. In general, the total concentration in ellagitannins is similar to that found in traditional oaks. In addition to ellagitannins, other hydrolysable tannins not present in oak have been found in chestnut. In addition, cherry and acacia have condensed tannins, never found in oak, in their composition. All this means that there is an important qualitative difference between these different woods and *Quercus* with respect to the composition of the traditional oak used in wine ageing. The main phenolic components analyzed specially in the green and seasoned wood of these new species of *Quercus* were ellagitannins (Table 1), with similar results to those found in other oaks traditionally used in oenology. Drying and toasting degrade these compounds as well as traditional *Quercus* woods. The eight ellagitannins identified in the traditional oaks were found in all the new species of *Quercus*, except in the case of the *Q. oocarpa*, which did not present dimer ellagitannins (A, B, C, and D Roburins) in its composition. *Q. pyrenaica*, *Q. faginea*, and *Q. oocarpa* had a similar range of total ellagitannin concentration among them; moreover, ellagitannins of these species were between *Q. robur* and *Q. alba*, and more similar to *Q. petraea*. However, *Q. humboldtii* showed a lower concentration than the other new species and more similar to *Q. alba*. On the other hand, *Q. frainetto* is distinguished especially from the other species by its higher content of pentosylated dimers, and a monomer concentration that is similar to the rest of *Quercus*, which makes it the species with the highest content in ellagitannins (108 mg/g). Unlike ellagitannins, low molecular weight compounds during drying and toasting increase their concentration in both *Quercus* and other woods, just like traditional oaks. *Castanea sativa* Mill. wood has the highest low molecular weight phenolic

(LMWP) content (Table 1). In general, the total LMWP contents found in the acacia (*Robinia pseudoacacia* L.), cherry (*Prunus*), and ash (*Fraxinus*) woods are lower than in the others, especially after drying. Ellagic and gallic acids are the main LMWPs in *Q. pyrenaica*, *Q. faginea*, *Q. humboldtii*, and *Castanea sativa* as in traditional oaks. Only in the toasted wood of *Q. humboldtii* does this not occur, as the majority were coniferaldehyde and sinapaldehyde. This behaviour has also been observed in traditional woods, especially in the species *Q. alba* and/or *Q. robur* [7,17,47,48], although the most common profile is that mentioned above. In *Q. oocarpa* and *Q. frainetto*, only the ellagic and gallic acids were studied [36], probably because they are the majority in this species. Furthermore, there is more ellagic acid than gallic acid in *Q. pyrenaica*, *Q. faginea*, *Q. humbolditti*, and *Q. oocarpa*, as usually occurs in traditional oaks; however, in *Q. frainetto* and *Castanea sativa*, this relationship was reversed. Compared to traditional oak, the most different woods were acacia, cherry, and ash, as in none of the three were these acids the majority. With the exception of the dry wood of *P. cereaus*, the majority component of which was ellagic acid, gallic acid was not detected [40]. Similarly, neither gallic acid nor ellagic acid were detected in the wood of the considered *Fraxinus* species [43].

Table 2 represents the concentration of some of the most representative volatile compounds of traditional oak for their aromatic contribution to wine during the ageing process analyzed by gas chromatography (GC) in green wood and dried and toasted specimens of the genus *Quercus*, but of different species to the traditional ones (*Quercus: pyrenaica, faginea,* and *humboldtti*) and woods other than oak (*castanea sativa, robina pseudoacacia, prunus,* and *fraxinus excelsior* and *american*). This table does not list *Q. oocarpa* and *Q. frainetto*, as no work has been found with the concentrations of these volatile compounds. Only Vivas [21] shows the presence of vanillin and eugenol β-Methyl-γ-octalactone in these species as well as *Q. alba, Q. petraea,* and *Q. robur* and oxo-3-retro-α-ionol in addition to *Q. alba*. Regarding cherry, in Table 2 the species is not indicated since it was not identified in the studies found. In addition, as these compounds are formed during drying and especially during toasting, only one work in green wood has been found that studied the species *Q. pyrenaica*. The concentration of the volatile species represented in Table 2 shows that the majority of the green and dried wood of *Q. pyrenaica* is cis-β-Methyl-γ-octalactone, 59 and 68 μg/g, respectively; in the dry wood of *Q. faginea* it is furfural with 17.6 μg/g, and in *Q. humbolditti* it is eugenol with 2.65 μg/g. However, the majority in non-*Quercus* woods after drying is vanillin. As for the wood after toasting, which is usually used in the processes of ageing wines, in almost all species the content of furfural is higher than that of the other five aromas as generally occurs in traditional oaks [17,51,55,60], with the exception of *Q. faginea*, *Fraxinus american*, and *Fraxinus excelsior*, with vanillin content being the highest. The guaiacol levels in medium-toasted ash woods were much higher than those detected in the other toasted woods, even those normally found in traditional oak, so a more pronounced smoke character can be expected when using toasted ash wood in ageing wines. In some studies, the concentrations of eugenol found in *Q. pyrenaica* were very high, especially when they were subjected to light toasting, much more so than in those normally found in traditional oaks, so that when using these woods we would have more spicy wines, especially with notes of clove. In general, Table 2 shows that woods not belonging to the genus *Quercus* do not have either β-Methyl-γ-octalactone or cis-β-Methyl-γ-octalactone in their composition. Only the study by Caldeira et al. [53] found small amounts of these isomers in chestnut (0.23 and 0.34 μg/g in the isomer trans and cis, respectively), but using 55% ethanol to extract them. The two isomers of β-Methyl-γ-octalactone have a high sensory impact on the wines after wood maturation [136], giving the wines coconut, toasted, and wood notes; moreover, theses isomers allow French and American oaks to be differentiated [59,137,138]. *Q. humboldtii* also presented very low concentrations of these two isomers. Regarding the species *Q. faginea*, the concentration of these two isomers in its wood has only been found in the work of Cadahía et al. [51], in which they also study the traditional oak species, observing that the concentrations in this species are within the usual values found in the traditional species (*Q. robur, Q. petraea,* and *Q. alba*). The wood of *Q. pyrenaica* has been more widely studied, so we find wider ranges of concentrations, depending on the origin, drying,

Beverages **2018**, *4*, 94

and toasting. In general, we could say that some woods have higher concentrations than those found in traditional woods, especially the cis isomer.

Many options for woods to be used in cooperage are available and suitable and there may be more. However, with the exception of the species *Q. pyrenaica*, there have not been many studies carried out on the aforementioned woods and the corresponding treatments in cooperage. Therefore, it is considered of great interest to know more about the aromatic composition of these woods, thus offering more information to coopers and oenologists about the wood they can use for their wines, thus providing that distinctive sought-after seal.

Author Contributions: Conceptualization, M.d.A.-S. and I.N.; Funding acquisition, I.N. and M.d.A.-S.; Methodology, I.N.; Supervision, M.d.A.-S.; Visualization, A.M.-G.; Writing—original draft, A.M.-G. and R.S.-G.; Writing—review & editing, A.M.-G., I.N. and M.d.A.-S.

Funding: This work was financed by Junta de Castilla y León (VA028U16), MINECO (AGL2014-54602-P), FEDER, and the Interreg España-Portugal Programme (Iberphenol). R.S.G. thanks Iberphenol for his contract.

Acknowledgments: The authors wish to thank Ann Holliday for her services in revising the English.

Conflicts of Interest: The authors declare no competing financial interest.

References

1. Gautier, J.F. Histoire et Actualité Du Tonneau. *Rev. Française D'oenologie* **2000**, *181*, 33–35.
2. Gautier, J.F. Le Tonneau à Travers Les âGes. *Rev. Oenologues Tech. Vitivinic. Oenologicques* **2003**, *30*, 13–15.
3. Vivas, N.; Saint-Cricq de Gaulejac, N. The useful lifespan of new barrels and risk related to the use of old barrels. *Aust. N. Z. Wine Ind. J.* **1999**, *14*, 37–45.
4. Singleton, V.L. Le Stockage Des Vins En Barriques: Utilisation et Variables Significatives. *J. Sci. Tech. Tonnellerie* **2000**, *6*, 1–25.
5. Fernández de Simón, B.; Cadahía, E. *Utilización Del Roble Español en el Envejecimiento de Vinos: Comparación con Roble Francés y Americano*; Instituto Nacional de Investigación y Tecnología Agraria y Alimentaria Ministerio de Educación y Ciencia: Madrid, Spain, 2004.
6. Chira, K.; Teissedre, P.-L. Chemical and sensory evaluation of wine matured in oak barrel: Effect of oak species involved and toasting process. *Eur. Food Res. Technol.* **2015**, *240*, 533–547. [CrossRef]
7. Cadahía, E.; Muñoz, L.; De Simón, B.F.; García-Vallejo, M.C. Changes in low molecular weight phenolic compounds in Spanish, French, and American oak woods during natural seasoning and toasting. *J. Agric. Food Chem.* **2001**, *49*, 1790–1798. [CrossRef] [PubMed]
8. Cadahía, E.; Varea, S.; Muñoz, L.; Fernández de Simón, B.; García-Vallejo, M.C. Evolution of ellagitannins in Spanish, French, and American oak woods during natural seasoning and toasting. *J. Agric. Food Chem.* **2001**, *49*, 3677–3684. [CrossRef] [PubMed]
9. Fernández de Simón, B.; Hernández, T.; Cadahía, E.; Dueñas, M.; Estrella, I. Phenolic compounds in a Spanish red wine aged in barrels made of Spanish, French and American oak wood. *Eur. Food Res. Technol.* **2003**, *216*, 150–156. [CrossRef]
10. Fernández De Simón, B.; Cadahía, E.; Jalocha, J. Volatile compounds in a Spanish red wine aged in barrels made of Spanish, French, and American oak wood. *J. Agric. Food Chem.* **2003**, *51*, 7671–7678. [CrossRef] [PubMed]
11. Cadahía, E.; Fernández de Simón, B.; Sanz, M.; Poveda, P.; Colio, J. Chemical and chromatic characteristics of Tempranillo, Cabernet Sauvignon and Merlot wines from DO Navarra aged in Spanish and French oak barrels. *Food Chem.* **2009**, *115*, 639–649. [CrossRef]
12. Guchu, E.; Díaz-Maroto, M.C.; Pérez-Coello, M.S.; González-Viñas, M.A.; Ibáñez, M.D.C. Volatile composition and sensory characteristics of chardonnay wines treated with American and Hungarian oak chips. *Food Chem.* **2006**, *99*, 350–359. [CrossRef]
13. Díaz-Maroto, M.C.; Guchu, E.; Castro-Vázquez, L.; de Torres, C.; Pérez-Coello, M.S. Aroma-active compounds of American, French, Hungarian and Russian oak woods, Studied by GC–MS and GC–O. *Flavour Fragr. J.* **2008**, *23*, 93–98. [CrossRef]
14. Mosedale, J.R.; Ford, A. Variation of the flavour and extractives of european oak wood from two French forests. *J. Sci. Food Agric.* **1996**, *70*, 273–287. [CrossRef]

15. Prida, A.; Puech, J.-L. Influence of geographical origin and botanical species on the content of extractives in American, French, and East European oak woods. *J. Agric. Food Chem.* **2006**, *54*, 8115–8126. [CrossRef] [PubMed]

16. Martínez-Gil, A.M.; Cadahía, E.; Fernández De Simón, B.; Gutiérrez-Gamboa, G.; Nevares, I.; Alamo-Sanza, M. *Quercus Humboldtii* (Colombian Oak): Characterization of oak heartwood phenolic composition with respect to traditional oak woods in oenology. *Ciência Técnica Vitivinícola* **2017**, *32*, 93–101. [CrossRef]

17. Martínez-Gil, A.; Cadahía, E.; Fernández de Simón, B.; Gutiérrez-Gamboa, G.; Nevares, I.; del Álamo-Sanza, M. Phenolic and volatile compounds in *Quercus Humboldtii* Bonpl. Wood: Effect of toasting with respect to oaks traditionally used in cooperage. *J. Sci. Food Agric.* **2018**. [CrossRef]

18. Singleton, V.L. Some aspects of the wooden container as a factor in wine maturation. *Chem. Winemak. Adv. Chem. Ser.* **1974**, *137*, 254–277.

19. Taransaud, J. *Le Livre de La Tonnellerie*; La Roue à Livres Diffusion: París, France, 1976.

20. Cazenave de la Roche, A. La Tonelería En El Contexto Marítimo de La Época Del Renacimiento: Estudio de Un Cargamento de Toneles Hallado En El Pecio de Villefranche s/Mer (1516). In *Actas de la XIVa Conferencia Nacional de Arqueología Argentina*; Universidad Nacional de Tucumán: Rosario, Argentina, 2004; pp. 1–17.

21. Vivas, N. *Manual De Tonelería: Destinado A Usuarios De Toneles*; Mundi, Pre: Madrid, Spain, 2005.

22. Feuillat, F.; Keller, R. Variability of Oak Wood (*Quercus Robur* L., *Quercus Petraea* Liebl.) Anatomy relating to cask properties. *Am. J. Enol. Vitic.* **1997**, *48*, 502–508.

23. Cadahía, E.; Fernández de Simón, B.; Poveda, P.; Sanz, M. *Utilización de Quercus Pyrenaica Willd. de Castilla y León En El Envejecimiento de Vinos. Comparación Con Roble Francés y Americano*; Instituto Nacional de Investigación y Tecnología Agraria y Alimentaria Ministerio de Educación y Ciencia: Madrid, Spain, 2008.

24. Timbal, J.; Kremer, A. Caractères Botaniques, Morphologiques et Clorologiques. In *Le chêne rouge d´Amérique*; Dremer, J.T., Le Goff, A., Nepveu, G., Eds.; INRA: París, France, 1994; pp. 45–53.

25. Fernández-Golfín, J.I.; Cadahía, E. *Características Físicas y Químicas de La Madera de Roble En La Fabricación de Barricas*; Gobierno de la Rioja: L. La Rioja, Spain, 1999; pp. 11–66.

26. Philp, J. Cask Quality and Warehouse Conditions. In *The Science and Technology of Whiskies*; Piggott, J.R., Sharp, R., Duncan, R.E.B., Eds.; Longman Scientific & Technical: Harlow, Essex, UK, 1989; pp. 264–294.

27. Fernández de Simón, B.; Sanz, M.; Cadahía, E.; Poveda, P.; Broto, M. Chemical characterization of oak heartwood from Spanish forests of *Quercus Pyrenaica* (Wild.). Ellagitannins, Low Molecular Weight Phenolic, and Volatile Compounds. *J. Agric. Food Chem.* **2006**, *54*, 8314–8321. [CrossRef] [PubMed]

28. Fernández de Simón, B.; Cadahía, E.; Conde, E.; García-Vallejo, M.C. Evolution of phenolic compounds of Spanish oak wood during natural seasoning. First Results. *J. Agric. Food Chem.* **1999**, *47*, 1687–1694. [CrossRef] [PubMed]

29. Fernández de Simón, B.; Cadahía, E.; Conde, E.; García-Vallejo, M.C. Ellagitannins in woods of Spanish, French and American oaks. *Holzforschung* **1999**, *53*, 147–150.

30. Fernández de Simón, B.; Cadahía, E.; Conde, E.; García-Vallejo, M.C. Low molecular weight phenolic compounds in Spanish oak woods. *J. Agric. Food Chem.* **1996**, *44*, 1507–1511. [CrossRef]

31. Alañón, M.E.; Castro-Vázquez, L.; Díaz-Maroto, M.C.; Hermosín-Gutiérrez, I.; Gordon, M.H.; Pérez-Coello, M.S. Antioxidant capacity and phenolic composition of different woods used in cooperage. *Food Chem.* **2011**, *129*, 1584–1590. [CrossRef]

32. Jordão, A.M.; Ricardo-Da-Silva, J.M.; Laureano, O. Ellagitannins from Portuguese Oak Wood (*Quercus Pyrenaica* Willd.) Used in Cooperage: Influence of geographical origin, coarseness of the grain and toasting level. *Holzforschung* **2007**, *61*, 155–160. [CrossRef]

33. Castro-Vázquez, L.; Alañón, M.E.; Ricardo-Da-Silva, J.M.; Pérez-Coello, M.S.; Laureano, O. Study of phenolic potential of seasoned and toasted Portuguese wood species (*Quercus Pyrenaica* and *Castanea Sativa*). *J. Int. des Sci. la Vigne du Vin* **2013**, *47*, 311–319. [CrossRef]

34. Canas, S.; Leandro, M.C.; Spranger, M.I.; Belchior, A.P. Influence of botanical species and geographical origin on the content of low molecular weight phenolic compounds of woods used in Portuguese cooperage. *Holzforschung* **2000**, *54*, 255–261. [CrossRef]

35. Canas, S.; Grazina, N.; Belchior, A.P.; Spranger, M.I.; de Sousa, R.B.; Sousa, R.B.D. Modelisation of heat treatment of Portuguese oak wood (*Quercus Pyrenaica* L.). Analysis of the behaviour of low molecular weight phenolic compounds. *Ciência Técnica Vitivinícola* **2000**, *15*, 75–94.

36. Vivas, N.; Glories, Y.; Bourgeois, G.; Vitry, C. The heartwood ellagitannins of different oaks (*Quercus* Sp.) and chestnut species (*Castanea Sativa* Mill.). Quantity analysis of red wines aging in barrels. *J. Sci. Tech. Tonnelerie* **1996**, *2*, 25–75.

37. Sanz, M.; Cadahía, E.; Esteruelas, E.; Muñoz, Á.M.; Fernández De Simón, B.; Hernández, T.; Estrella, I. Phenolic compounds in chestnut (*Castanea Sativa* Mill.) heartwood. Effect of toasting at cooperage. *J. Agric. Food Chem.* **2010**, *58*, 9631–9640. [CrossRef] [PubMed]

38. Viriot, C.; Scalbert, A.; Hervé du Penhoat, C.L.M.; Moutounet, M. Ellagitannins in woods of sessile oak and sweet chestnut dimerization and hydrolysis during wood ageing. *Phytochemistry* **1994**, *36*, 1253–1260. [CrossRef]

39. Canas, S.; Leandro, M.C.; Spranger, M.I.; Belchior, A.P. Low molecular weight organic compounds of chestnut wood (*Castanea Sativa* L.) and corresponding aged brandies. *J. Agric. Food Chem.* **1999**, *47*, 5023–5030. [CrossRef] [PubMed]

40. Soares, B.; Garcia, R.; Freitas, A.M.C.; Cabrita, M.J. Phenolic compounds released from oak, cherry, chestnut and robinia chips into a syntethic wine: Influence of toasting level. *Ciência Técnica Vitivinic* **2012**, *27*, 17–26.

41. Canas, S.; Caldeira, I.; Mateus, A.M.; Belchior, A.P.; Clímaco, M.C.; Bruno-de-Sousa, R. Effect of natural seasoning on the chemical composition of chestnut wood used for barrel making. *Ciência Técnica Vitivinic* **2006**, *21*, 1–16.

42. Sanz, M.; Fernández de Simón, B.; Esteruelas, E.; Muñoz, A.M.; Cadahía, E.; Hernández, T.; Estrella, I.; Pinto, E. Effect of toasting intensity at cooperage on phenolic compounds in acacia (*Robinia Pseudoacacia*) heartwood. *J. Agric. Food Chem.* **2011**, *59*, 3135–3145. [CrossRef] [PubMed]

43. Sanz, M.; Fernández de Simón, B.; Cadahía, E.; Esteruelas, E.; Muñoz, Á.M.; Teresa Hernández, M.; Estrella, I. Polyphenolic profile as a useful tool to identify the wood used in wine aging. *Anal. Chim. Acta* **2012**, *732*, 33–45. [CrossRef] [PubMed]

44. Sanz, M.; Cadahía, E.; Esteruelas, E.; Muñoz, M.; Fernández De Simón, B.; Hernández, T.; Estrella, I. Phenolic compounds in cherry (*Prunus avium*) heartwood with a view to their use in cooperage. *J. Agric. Food Chem.* **2010**, *58*, 4907–4914. [CrossRef] [PubMed]

45. Sanz, M.; De Simón, B.F.; Cadahía, E.; Esteruelas, E.; Muñoz, A.M.; Hernández, T.; Estrella, I.; Pinto, E. LC-DAD/ESI-MS/MS study of phenolic compounds in ash (*Fraxinus Excelsior* L. and *F. Americana* L.) heartwood. effect of toasting intensity at cooperage. *J. Mass Spectrom.* **2012**, *47*, 905–918. [CrossRef] [PubMed]

46. Jordão, A.M.; Lozano, V.; Correia, A.C.; Ortega-Heras, M.; González-SanJosé, M.L. Comparative analysis of volatile and phenolic composition of alternative wood chips from cherry, acacia and oak for potential use in enology. *BIO Web Conf.* **2016**, *7*, 02012. [CrossRef]

47. Canas, S.; Belchior, A.P.; Falcão, A.; Gonçalves, J.A.; Spranger, M.I.; Bruno-De-Sousa, R. Effect of heat treatment on the thermal and chemical modifications of oak and chestnut wood used in brandy ageing. *Ciência Técnica Vitivinic* **2007**, *22*, 5–14.

48. Madrera, R.R.; Valles, B.S.; García, Y.D.; Argüelles, P.D.V.; Lobo, A.P. Alternative woods for aging distillates -an insight into their phenolic profiles and antioxidant activities. *Food Sci. Biotechnol.* **2010**, *19*, 1129–1134. [CrossRef]

49. Sanz, M.; Fernández de Simón, B.; Esteruelas, E.; Muñoz, Á.M.; Cadahía, E.; Teresa Hernández, M.; Estrella, I.; Martinez, J. Polyphenols in red wine aged in acacia (*Robinia Pseudoacacia*) and oak (*Quercus Petraea*) wood barrels. *Anal. Chim. Acta* **2012**, *732*, 83–90. [CrossRef] [PubMed]

50. Cadahía, E.; De Simón, B.F.; Vallejo, R.; Sanz, M.; Broto, M. Volatile compound evolution in Spanish oak wood (*Quercus Petraea* and *Quercus Pyrenaica*) during natural seasoning. *Am. J. Enol. Vitic.* **2007**, *58*, 163–172.

51. Cadahía, E.; Fernández de Simón, B.; Jalocha, J. Volatile compounds in Spanish, French, and American oak woods after natural seasoning and toasting. *J. Agric. Food Chem.* **2003**, *51*, 5923–5932. [CrossRef] [PubMed]

52. Alañón, M.E.; Pérez-Coello, M.S.; Díaz-Maroto, I.J.; Martín-Alvarez, P.J.; Vila-Lameiro, P.; Díaz-Maroto, M.C. Influence of geographical location, site and silvicultural parameters, on volatile composition of *Quercus Pyrenaica* Willd. Wood used in wine aging. *For. Ecol. Manag.* **2011**, *262*, 124–130. [CrossRef]

53. Caldeira, I.; Clímaco, M.C.; Bruno De Sousa, R.; Belchior, A.P. Volatile composition of oak and chestnut woods used in brandy ageing: Modification induced by heat treatment. *J. Food Eng.* **2006**, *76*, 202–211. [CrossRef]

54. Jordao, A.M.; Ricardo-Da-Silva, J.M.; Laureano, O. Comparison of volatile composition of cooperage oak wood of different origins (*Quercus Pyrenaica* vs. *Quercus Alba* and *Quercus Petraea*). *Mitteilungen Klosterneubg.* **2005**, *55*, 22–31.
55. De Simon, B.F.; Esteruelas, E.; Muñoz, À.M.; Cadahía, E.; Sanz, M. Volatile compounds in acacia, chestnut, cherry, ash, and oak woods, with a view to their use in cooperage. *J. Agric. Food Chem.* **2009**, *57*, 3217–3227. [CrossRef] [PubMed]
56. Flamini, R.; Dalla Vedova, A.; Cancian, D.; Panighel, A.; De Rosso, M. GC/MS-positive ion chemical ionization and MS/MS study of volatile benzene compounds in five different woods used in barrel making. *J. Mass Spectrom.* **2007**, *42*, 641–646. [CrossRef] [PubMed]
57. Fernández De Simõn, B.; Sanz, M.; Cadahía, E.; Esteruelas, E.; Muñoz, A.M. Nontargeted GC-MS approach for volatile profile of toasting in cherry, chestnut, false acacia, and ash wood. *J. Mass Spectrom.* **2014**, *49*, 353–370. [CrossRef] [PubMed]
58. Fernández de Simón, B.; Cadahía, E.; del Álamo, M.; Nevares, I. Effect of size, seasoning and toasting in the volatile compounds in toasted oak wood and in a red wine treated with them. *Anal. Chim. Acta* **2010**, *660*, 211–220. [CrossRef] [PubMed]
59. Fernández De Simón, B.; Muiño, I.; Cadahía, E. Characterization of volatile constituents in commercial oak wood chips. *J. Agric. Food Chem.* **2010**, *58*, 9587–9596. [CrossRef] [PubMed]
60. Fernández de Simón, B.; Cadahía, E.; Muiño, I.; del Álamo, M.; Nevares, I. Volatile composition of toasted oak chips and staves and of red wine aged with them. *Am. J. Enol. Vitic.* **2010**, *61*, 157–165.
61. III-IFE (Inventario Forestal Español) (Ed.) *Ministerio de Medio Ambiente*; Direccíon General de Conservación de la Naturaleza: Madrid, Spain, 2002.
62. Caudullo, G.; Welk, E.; San-Miguel-Ayanz, J. Chorological Maps for the Main European Woody Species. *Data Brief* **2017**, *12*, 662–666. [CrossRef] [PubMed]
63. Fernández de Simón, B.; Cadahía, E.; Conde, E.; García-Vallejo, M.C. Ellagitannins in woods of Spanish oaks. *J. Sci. Tech. Tonnellerie* **1998**, *4*, 91–97.
64. Fernández de Simón, B.; Cadahía, E.; Conde, E.; García-Vallejo, M.C. Low molecular weight phenolic compounds in woods of Spanish, French and American oak. *J. Sci. Tech. Tonnelerie* **1996**.
65. Jordão, A.M.; Ricardo-Da-Silva, J.M.; Laureano, O.; Adams, A.; Demyttenaere, J.; Verhé, R.; De Kimpe, N. Volatile composition analysis by solid-phase microextraction applied to oak wood used in cooperage (*Quercus Pyrenaica* and *Quercus Petraea*): Effect of botanical species and toasting process. *J. Wood Sci.* **2006**, *52*, 514–521. [CrossRef]
66. Canas, S. Phenolic composition and related properties of aged wine spirits: Influence of barrel characteristics. A Review. *Beverages* **2017**, *3*, 55. [CrossRef]
67. Canas, S.; Silva, V.; Belchior, A.P. Wood related chemical markers of aged wine brandies. *Ciência Técnica Vitivinícola* **2008**, *23*, 45–52.
68. Caldeira, I.; Belchior, A.P.; Clímaco, M.C.; Bruno De Sousa, R. Aroma profile of Portuguese brandies aged in chestnut and oak woods. *Anal. Chim. Acta* **2002**, *458*, 55–62. [CrossRef]
69. Caldeira, I.; de Sousa, R.B.; Belchior, A.P.; Climaco, M.C. A Sensory and chemical approach to the aroma of wooden agend lourinha wine brandy. *Ciencia Tecnica* **2008**, *23*, 97–110.
70. Carvalho, E.; Belchior, A.P.; Costa, S.; Caldeira, I.; Tralhao, I. Incidência Da Origem e Queima Da Madeira de Carvalho (*"Q. Pyrenaica, Q. Robur, Q. Sessiliflora, Q. Alba/Q. Stellata + Q. Lyrata/Q.Bicolor"*) e de Castanho (*"C. Sativa"*) Em Características Físico-Químicas e Organolépticas de Aguardentes Lourinha Em Envelhecim. *Ciência Técnica Vitivinícola* **1998**, *13*, 107–119.
71. Tavares, M.; Jordão, A.M.; Ricardo-Da-Silva, J.M. Impact of cherry, acacia and oak chips on red wine phenolic parameters and sensory profile. *OENO ONE* **2017**, *51*, 329–342. [CrossRef]
72. Sánchez-Gómez, R.; Nevares, I.; Martínez-Gil, A.; del Alamo-Sanza, M. Oxygen consumption by red wines under different micro-oxygenation strategies and *Q. Pyrenaica* chips. Effects on color and phenolic characteristics. *Beverages* **2018**, *4*, 69. [CrossRef]
73. Del Álamo, M.; Nevares, I.; Gallego, L.; Fernández de Simón, B.; Cadahía, E. Micro-oxygenation strategy depends on origin and size of oak chips or staves during accelerated red wine aging. *Anal. Chim. Acta* **2010**, *660*, 92–101. [CrossRef] [PubMed]

74. Rodríguez-Bencomo, J.J.; Ortega-Heras, M.; Pérez-Magariño, S.; González-Huerta, C. Volatile compounds of red wines macerated with Spanish, American, and French oak chips. *J. Agric. Food Chem.* **2009**, *57*, 6383–6391. [CrossRef] [PubMed]

75. Gonalves, F.J.; Jordao, A.M. Changes in antioxidant activity and the proanthocyanidin fraction of red wine aged in contact with Portuguese (*Quercus Pyrenaica* Willd.) and American (*Quercus Alba* L.) oak wood chips. *Ital. J. Food Sci.* **2009**, *21*, 51–64.

76. De Simón, B.F.; Cadahía, E.; Sanz, M.; Poveda, P.; Perez-Magariño, S.; Ortega-Heras, M.; González-Huerta, C. Volatile compounds and sensorial characterization of wines from four Spanish denominations of origin, aged in Spanish rebollo (*Quercus Pyrenaica* Willd.) oak wood barrels. *J. Agric. Food Chem.* **2008**, *56*, 9046–9055. [CrossRef] [PubMed]

77. Coninck, G.D.E.; Jordão, A.M.; Ricardo-Da-Silva, J.M.; Laureano, O. Evolution of phenolic composition and sensory properties in red wine aged in contact with Portuguese. *J. Int. des Sci. de la vigne et du vin* **2006**, *40*, 25–34.

78. Fernández De Simón, B.; Cadahía, E.; Hernández, T.; Estrella, I. Evolution of oak-related volatile compounds in a Spanish red wine during 2 years bottled, after aging in barrels made of Spanish, French and American oak wood. *Anal. Chim. Acta* **2006**, *563*, 198–203. [CrossRef]

79. Gallego, L.; Del Alamo, M.; Nevares, I.; Fernández, J.A.; De Simón, B.F.; Cadahía, E. Phenolic Compounds and Sensorial Characterization of Wines Aged with Alternative to Barrel Products Made of Spanish Oak Wood (*Quercus Pyrenaica* Willd.). *Food Sci. Technol. Int.* **2012**, *18*, 151–165. [CrossRef] [PubMed]

80. Miranda, I.; Sousa, V.; Ferreira, J.; Pereira, H. Chemical characterization and extractives composition of heartwood and sapwood from *Quercus Faginea*. *PLoS ONE* **2017**, *12*, 1–14. [CrossRef] [PubMed]

81. Ministerio de Medioambiente. *Plan Forestal Español*; Ministerio de Medioambiente: Madrid, Spain, 2002.

82. Reboredo, F.; Pais, J. Evolution of forest cover in Portugal: A review of the 12th–20th centuries. *J. For. Res.* **2014**, *25*, 249–256. [CrossRef]

83. Mauri, A.; Enescu, C.M.; Houston Durrant, T.; de Rigo, D.; Caudullo, G. Quercus Frainetto in Europe: Distribution, Habitat, Usage and Threats. In *European Atlas of Forest Tree Species*; San-Miguel-Ayanz, J., de Rigo, D., Caudullo, G., Houston Durrant, T., Mauri, A., Eds.; Publication Office of the European Union: Luxembourg, 2016; pp. 1–78.

84. Madrigal-Jiménez, T.A. Fenología y Ecofisiología Del Quercus Oocarpa (Fagaceae), Cartago, Costa Rica. *Rev. Biol. Trop.* **1996**, *44*, 117–123.

85. Andrés, A.M.; Luis Mario, C.C. Conservation and Sustainable Use of Oak Forests in the Conservation Corridor Guantiva-La Rusia—Iguaque, Santander and Boyac, Colombia. *Colomb. For.* **2010**, *13*, 5–30.

86. Argoti, J.C.; Salido, S.; Linares-Palomino, P.J.; Ramírez, B.; Insuasty, B.; Altarejos, J. Antioxidant activity and free radical-scavenging capacity of a selection of wild-growing colombian plants. *J. Sci. Food Agric.* **2011**, *91*, 2399–2406. [CrossRef] [PubMed]

87. González, R.E.; Baleta, L.C. Quantification and comparison of ageing markers substances of accelerated aging rums and in oak (*Quercus Humboldtii* Bonpland) barrels. *Rev. Venez. Cienc. Tecnol. Aliment.* **2010**, *1*, 170–183.

88. González, R.E.; Calderón, L.S.; Cabeza, R.A. Quantification of aging markers substances in *Quercus Humboldtii* through high efficiency liquid chromatography. *Temas Agrar.* **2008**, *13*, 56–63. [CrossRef]

89. Martínez-Gil, A.M.; del Alamo-Sanza, M.; Gutiérrez-Gamboa, G.; Moreno-Simunovic, Y.; Nevares, I. Volatile composition and sensory characteristics of Carménère wines macerating with Colombian (*Quercus Humboldtii*) oak chips compared to wines macerated with American (*Q. Alba*) and European (*Q. Petraea*) oak chips. *Food Chem.* **2018**, *266*, 90–100. [CrossRef] [PubMed]

90. Gortzi, O.; Metaxa, X.; Mantanis, G.; Lalas, S. Effect of artificial ageing using different wood chips on the antioxidant activity, resveratrol and catechin concentration, sensory properties and colour of two greek red wines. *Food Chem.* **2013**, *141*, 2887–2895. [CrossRef] [PubMed]

91. Canas, S.; Caldeira, I.; Belchior, A.P. Extraction/Oxidation kinetics of low molecular weight compounds in wine brandy resulting from different ageing technologies. *Food Chem.* **2013**, *138*, 2460–2467. [CrossRef] [PubMed]

92. De Rosso, M.; Cancian, D.; Panighel, A.; Dalla Vedova, A.; Flamini, R. Chemical compounds released from five different woods used to make barrels for aging wines and spirits: Volatile compounds and polyphenols. *Wood Sci. Technol.* **2009**, *43*, 375–385. [CrossRef]

93. Peng, S.; Scalbert, A.; Monties, B. Insoluble ellagitannins in castanea sativa and quercus petraea woods. *Phytochemistry* **1991**, *30*, 775–778. [CrossRef]

94. Caldeira, I.; Anjos, O.; Belchior, A.P.; Canas, S. Sensory Impact of alternative ageing technology for the production of wine brandies. *Ciência Técnica Vitivinícola* **2017**, *32*, 12–22. [CrossRef]

95. Caldeira, I.; Santos, R.; Ricardo-Da-Silva, J.M.; Anjos, O.; Mira, H.; Belchior, A.P.; Canas, S. Kinetics of odorant compounds in wine brandies aged in different systems. *Food Chem.* **2016**, *211*, 937–946. [CrossRef] [PubMed]

96. Anjos, O.; Carmona, C.; Caldeira, I.; Canas, S. Variation of extractable compounds and lignin contents in wood fragments used in the aging of wine brandies. *BioResources* **2013**, *8*, 4484–4496. [CrossRef]

97. Caldeira, I.; Belchior, A.P.; Canas, S. Effect of alternative ageing systems on the wine brandy sensory profile. *Ciência Técnica Vitivinícola* **2013**, *28*, 9–18.

98. Caldeira, I.; Anjos, O.; Portal, V.; Belchior, A.P.; Canas, S. Sensory and chemical modifications of wine-brandy aged with chestnut and oak wood fragments in comparison to wooden barrels. *Anal. Chim. Acta* **2010**, *660*, 43–52. [CrossRef] [PubMed]

99. Canas, S.; Caldeira, I.; Belchior, A.P. Comparison of alternative systems for the ageing of wine brandy. Oxygenation and wood shape effect. *Ciência Técnica Vitivinícola* **2009**, *24*, 91–99.

100. Canas, S.; Caldeira, I.; Belchior, A.P.; Spranger, M.I.; Clímaco, M.C.; Bruno-de-Sousa, R. *Chestnut Wooden Barrels for the Ageing of Wine Spirits*; OIV: París, France, 2018; pp. 1–16.

101. Canas, S.; Belchior, A.P.; Mateus, A.M.; Spranger, M.I.; Bruno-de-Sousa, R. Kinetics of impregnation/evaporation and release of phenolic compounds from wood to brandy in experimental model. *Ciência Técnica Vitivinícola* **2002**, *17*, 1–14.

102. Belchior, A.P.; Caldeira, I.; Costa, S.; Lopes, C.; Tralhão, G.; Ferrão, A.F.M.; Mateus, A.M.; Carvalho, E. Evolução Das Características Fisico-Químicas e Organolépticas de Aguardentes Lourinhã Ao Longo de Cinco Anos de Envelhecimento Em Madeiras de Carvalho e de Castanheiro. *Ciência Técnica Vitivinícola* **2001**, *16*, 81–94.

103. Canas, S.; Caldeira, I.; Anjos, O.; Lino, J.; Soares, A.; Pedro Belchior, A. Physicochemical and sensory evaluation of wine brandies aged using oak and chestnut wood simultaneously in wooden barrels and in stainless steel tanks with staves. *Int. J. Food Sci. Technol.* **2016**, *51*, 2537–2545. [CrossRef]

104. Cerezo, A.B.; Tesfaye, W.; Torija, M.J.; Mateo, E.; García-Parrilla, M.C.; Troncoso, A.M. The phenolic composition of red wine vinegar produced in barrels made from different woods. *Food Chem.* **2008**, *109*, 606–615. [CrossRef]

105. Callejón, R.M.; Torija, M.J.; Mas, A.; Morales, M.L.; Troncoso, A.M. Changes of volatile compounds in wine vinegars during their elaboration in barrels made from different woods. *Food Chem.* **2010**, *120*, 561–571. [CrossRef]

106. Arfelli, G.; Sartini, E.; Corzani, C.; Fabiani, A.; Natali, N. Impact of wooden barrel storage on the volatile composition and sensorial profile of red wine. *Food Sci. Technol. Int.* **2007**, *13*, 293–299. [CrossRef]

107. Gambuti, A.; Capuano, R.; Lisanti, M.T.; Strollo, D.; Moio, L. Effect of aging in new oak, one-year-used oak, chestnut barrels and bottle on color, phenolics and gustative profile of three monovarietal red wines. *Eur. Food Res. Technol.* **2010**, *231*, 455–465. [CrossRef]

108. Alañón, M.E.; Schumacher, R.; Castro-Vázquez, L.; Díaz-Maroto, I.J.; Díaz-Maroto, M.C.; Pérez-Coello, M.S. Enological potential of chestnut wood for aging Tempranillo wines part I: Volatile compounds and sensorial properties. *Food Res. Int.* **2013**, *51*, 325–334. [CrossRef]

109. Alañón, M.E.; Schumacher, R.; Castro-Vázquez, L.; Díaz-Maroto, M.C.; Hermosín-Gutiérrez, I.; Pérez-Coello, M.S. Enological potential of chestnut wood for aging Tempranillo wines part II: Phenolic compounds and chromatic characteristics. *Food Res. Int.* **2013**, *51*, 536–543. [CrossRef]

110. Fernández De Simón, B.; Sanz, M.; Cadahía, E.; Martínez, J.; Esteruelas, E.; Muñoz, A.M.; De Simón, B.F.; Sanz, M.; Cadahía, E.; Martínez, J.; et al. Polyphenolic compounds as chemical markers of wine ageing in contact with cherry, chestnut, false acacia, ash and oak wood. *Food Chem.* **2014**, *143*, 66–76. [CrossRef] [PubMed]

111. Fernández De Simón, B.; Martínez, J.; Sanz, M.; Cadahía, E.; Esteruelas, E.; Muñoz, A.M. Volatile compounds and sensorial characterisation of red wine aged in cherry, chestnut, false acacia, ash and oak wood barrels. *Food Chem.* **2014**, *147*, 346–356. [CrossRef] [PubMed]

112. Palomero, F.; Bertani, P.; Fernández De Simón, B.; Cadahía, E.; Benito, S.; Morata, A.; Suárez-Lepe, J.A. Wood impregnation of yeast lees for winemaking. *Food Chem.* **2015**, *171*, 212–223. [CrossRef] [PubMed]
113. De Rosso, M.; Panighel, A.; Dalla Vedova, A.; Stella, L.; Flamini, R. Changes in chemical composition of a red wine aged in acacia, cherry, chestnut, mulberry, and oak wood barrels. *J. Agric. Food Chem.* **2009**, *57*, 1915–1920. [CrossRef] [PubMed]
114. Keil, G.; Spavento, E.; Murace, M.; Millanes, A. The influence of natural seasoning on the concentration of eugenol, vanillin and cis and trans-β-methyl-γ-octalactone extracted from French and American oak wood. *For. Syst.* **2011**, *20*, 21–26.
115. Citron, G. Uso Del Legno in Enologia: Specie Botaniche Utiliz- Zate. Anatomia e Classifi Cazione. *L'Informatore Agrar.* **2005**, *59*, 69–72.
116. Kozlovic, G.; Jeromel, A.; Maslov, L.; Pollnitz, A.; Orlić, S. Use of acacia barrique barrels—Influence on the quality of malvazija from Istria wines. *Food Chem.* **2010**, *120*, 698–702. [CrossRef]
117. Psarra, C.; Gortzi, O.; Makris, D.P. Kinetics of polyphenol extraction from wood chips in wine model solutions: Effect of chip amount and botanical species. *J. Inst. Brew.* **2015**, *121*, 207–212. [CrossRef]
118. Alañón, M.E.; Marchante, L.; Alarcón, M.; Díaz-Maroto, I.J.; Pérez-Coello, S.; Díaz-Maroto, M.C. Fingerprints of acacia aging treatments by barrels or chips based on volatile profile, sensorial properties, and multivariate analysis. *J. Sci. Food Agric.* **2018**. [CrossRef] [PubMed]
119. Delia, L.; Jordao, A.M.; Ricardo-Da-Silva, J.M. Influence of different wood chips species (oak, acacia and cherry) used in a short period of aging on the quality of encruzado white wines. *Mitteilungen Klosterneuburg* **2017**, *67*, 84–96.
120. Cerezo, A.B.; Espartero, J.L.; Winterhalter, P.; García-Parrilla, M.C.; Troncoso, A.M. (+)-Dihydrorobinetin: A marker of vinegar aging in acacia (Robinia pseudoacacia) wood. *J. Agric. Food Chem.* **2009**, *57*, 9551–9554. [CrossRef] [PubMed]
121. Chatonnet, P.; Dubourdie, D.; Boidron, J.; Pons, M. The Origin of ethylphenols in wines. *J. Sci. Food Agric.* **1992**, *60*, 165–178. [CrossRef]
122. Torija, M.-J.; Mateo, E.; Vegas, C.-A.; Jara, C.; González, Á.; Poblet, M.; Reguant, C.; Guillamon, J.-M.; Mas, A. Effect of wood type and thickness on acetification kinetics in traditional vinegar production. *Int. J. Wine Res.* **2009**, *1*, 155–160.
123. Elbert, L.; Little, J. *USGS Geosciences and Environmental Change Science Center: Digital Representations of Tree Species Range Maps from "Atlas of United States Trees"*; Elbert, L., Little, J., Eds.; U.S. Department of the Interior/U.S. Geological Survey: Lakewood, CO, USA, 2016.
124. Culleré, L.; Fernández de Simón, B.; Cadahía, E.; Ferreira, V.; Hernández-Orte, P.; Cacho, J. Characterization by gas chromatography-olfactometry of the most odor-active compounds in extracts prepared from acacia, chestnut, cherry, ash and oak woods. *LWT Food Sci. Technol.* **2013**, *53*, 240–248. [CrossRef]
125. Chinnici, F.; Natali, N.; Sonni, F.; Bellachioma, A.; Riponi, C. Comparative changes in color features and pigment composition of red wines aged in oak and cherry wood casks. *J. Agric. Food Chem.* **2011**, *59*, 6575–6582. [CrossRef] [PubMed]
126. Chinnici, F.; Natali, N.; Bellachioma, A.; Versari, A.; Riponi, C. Changes in phenolic composition of red wines aged in cherry wood. *LWT Food Sci. Technol.* **2015**, *60*, 977–984. [CrossRef]
127. Prieto, R.O.; González-Oliva, L. Lista Nacional de Plantas Invasoras y Potencialmente Invasoras En La República de Cuba. *Bissea Boletín Sobre Conserv. Plantas del Jardín Botánico Nac. Cuba* **2015**, *9*, 5–90.
128. Culiță, S.; Adrian, O.; Pavol Jun, E.; Peter, F. New contribution to the study of alien flora in Romania. *J. Plant Dev.* **2011**, *18*, 121–134.
129. Azizian, D.; Azizian, D. Morphology and distribution of trichomes in some genera (Morus, Ficus, Broussonetia and Maclura) of Moraceae. *Iran. J. Bot.* **2002**, *9*, 195–202.
130. Nepal, M.; Ferguson, C.J.; Carolyn, J. Phylogenetics of Morus (Moraceae) inferred from ITS and TrnL-TrnF sequence data. *Syst. Bot.* **2012**, *37*, 442–450. [CrossRef]
131. Koek-Noorman, J.; Topper, S.M.C.; ter Welle, B.J.H. The systematic wood anatomy of the Moraceae (Urticales) V. Genera of the tribe moreae without urticaceous stamens. *IAWA J.* **1986**, *7*, 175–193.
132. Watanabe, T. Substances in mulberry leaves which attract silkworm larvae (*Bombyx mori* L.). *Nature* **1958**, *182*, 325–326. [CrossRef]
133. Karami, E.; Pourtahmasi, K.; Shahverdi, M. Wood anatomical structure of *Morus alba* L. and *Morus nigra* L., native to Iran. *Not. Sci. Biol.* **2010**, *2*, 129–132. [CrossRef]

134. Pasheva, M.; Nashar, M.; Pavlov, D.; Slavova, S.; Ivanov, D.; Ivanova, D. Antioxidant Capacity of Different Woods Traditionally Used for Coloring Hard Alcoholic Beverages in Bulgaria. *Sci. Technol.* **2013**, *3*, 123–127.
135. Chatonnet, P.; Boidron, J.N. Influence Du Traitement Thermique Du Bois de Chine Sur Sa Composition Chimique 1 e Partie: Définition Des Parametres Thermiques de La Chauffe Des Fûts En Tonnellerie. *Connaiss. Vigne Vin* **1989**, *23*, 1–11.
136. Boidron, J.N.; Chatonnet, P.; Pons, M. Influence Du Bois Sur Certaines Substances Odorantes Des Vins. *Connaiss. Vigne Vin* **1988**, *22*, 275. [CrossRef]
137. Sefton, M.A.; Francis, I.L.; Pocock, K.F.; Williams, P. The Influence of natural seasoning on the concentration of eugenol, vanillin and cis and trans-β-methyl-γ-octalactone extracted from French and American oak wood. *Sci. Aliments* **1993**, *13*, 629–644.
138. Waterhouse, A.L.; Towey, J.P. Oak lactone isomer ratio distinguishes between wine fermented in american and french oak barrels. *J. Agric. Food Chem.* **1994**, *42*, 1971–1974. [CrossRef]

Article

Study of High Power Ultrasound for Oak Wood Barrel Regeneration: Impact on Wood Properties and Sanitation Effect

Marion Breniaux [1],*, Philippe Renault [1,2], Fabrice Meunier [3] and Rémy Ghidossi [1]

[1] Université Bordeaux, ISVV, EA 4577, Unité de recherche Œnologie, F-33882 Villenave d'Ornon, France; philippe.renault33@gmail.com (P.R.); remy.ghidossi@u-bordeaux.fr (R.G.)

[2] Dyogena, 33290 Blanquefort, France

[3] Amarante Process, ADERA, 33600 Pessac, France; fabrice.meunier@u-bordeaux.fr

* Correspondence: marion.breniaux@u-bordeaux.fr; Tel.: +0033-5-57-57-58-58

Received: 21 December 2018; Accepted: 16 January 2019; Published: 1 February 2019

Abstract: This study aims to investigate the ability of high power ultrasound (HPU) to ensure oak barrel sterilization and wood structure preservation. Optimization was performed in terms of temperature and time and the impact of the HPU process on the porous material was also characterized. In this research, several wood characteristics were considered, such as the specific surface area, hydrophobicity, oxygen desorption and spoilage microorganisms after treatment. The study showed that the microbial stabilization could be obtained with HPU 60 °C/6 min. The results obtained show that microorganisms are impacted up to a depth of 9 mm, with a *Brettanomyces bruxellensis* population < 1 log CFU/g. The operating parameters used during the HPU treatment can also impact on wood exchange surface and oxygen desorption kinetics indicating that tartrate is removed. Indeed, the total oxygen desorption rate was recovered after HPU treatment, close to a new oak barrel, and thus may indicate that there is no impact on the ultrastructure (vessel, pore size or rays). Finally, wood wettability can also be impacted, depending on the temperature and the duration of exposure.

Keywords: high power ultrasound; wine aging; regeneration; sanitation; *brettanomyces*; oak wood barrel

1. Introduction

Aging red wines can be carried out in barrels to allow olfactory and gustatory modifications [1–4]. However, when aging takes place in wooden barrels, some organoleptic deviations may appear referred to as "brett flavor", referring to "stable", "leather", "manure" or "horse sweat". Bacteria and yeasts can penetrate into the wood to the same depth as the wine (close to 8 mm) [5] and then contaminate other wines if the oak barrels are not sterilized properly [6], even if polyphenols have a negative impact on bacteria viability [7]. *Brettanomyces* can provide organoleptic deviations due to production of undesirable compounds, such as 4-ethylphenol and 4-ethylgaiacol [8]. These molecules can deteriorate the wine quality [9]. To avoid such deviations, several treatments can be employed in wineries. However, Yap et al. [10] have argued that hot water and chemical treatments (the two treatments most commonly used) may be ineffective against the *Brettanomyces* spoilage yeast [11,12]. Several types of treatments are used in wineries to sanitize barrels, such as chemical agents (sulfur dioxide, ozone) or physical agents, with UV radiation, hot water, microwaves and ultrasounds showing varying levels of efficiency [5,11,13]. For example, microwave treatment enables only a 35% reduction in the *Brettanomyces* spp. population [14]. Guzzon et al. [13] studied the efficacy of aqueous steam, UV irradiation, gaseous, and aqueous O_3 for sterilizing oak barrels. Steam and O_3 were demonstrated to be the most effective treatments, eliminating as much as 90% of yeasts. Nevertheless, in this case, total sanitation is not completely ensured because the treatment is only carried out on the stave surface. Due to the porous nature of wood and the limited transmittance of UV radiation, this process appears to be ineffective. With regards to the steam treatment, some authors have shown that the microorganism inhibition is linked to the wood depth. A reduction of 3 log can be observed for depths of less than 3 mm, and 2 log if the depth is between 3 and 6 mm [15]. None of these processes are therefore completely effective and sulfur dioxide in wines should be managed carefully to avoid contamination.

Studies of the use of high power ultrasound (HPU) in industrial processes has recently been published [16,17]. In this process, electrical energy can be converted into ultrasound (20 kHz–10 MHz) at frequencies higher than those audible by the human ear (16–20 kHz). High power ultrasound is characterized by intensities in excess of 1 W/cm^2 and frequencies between 20 and 100 kHz [18–20]. When they are emitted in a liquid, this process forms high-energy micro bubbles (acoustic cavitation phenomenon) [21–23]. These cavitation bubbles generate very high temperatures locally (close to 80 °C) and pressures greater than 50 MPa [19]. These micro cavitation bubbles (diameters around 1 μm) act homogeneously throughout the fluid (Pascal's law) and can penetrate deeply into the pores of the wood. Piyasena et al. [24] showed that the cavitation phenomenon generated by HPU can kill microbial cells by cell disintegration in many cases [24]. The effect of ultrasound on the growth and viability of pathogenic bacteria such as *Escherichia coli*, *Pseudomonas fluorescens*, several yeast like *Saccharomyces cerevisiae*, *Brettanomyces*, as well as various fungi, algae and protozoa, has been summarized by Jiranek et al. [25]. These authors consider that the rate of inactivation of microorganisms varies with the power [26–28] and frequency [29] of the ultrasound applied. Thus, according to Tsukamoto et al. [28], the wave amplitude has a large influence on the inactivation rate for *S. cerevisiae* cells. According to Borthwick et al. [29], cell disruption in this species is greater at high frequencies (267 kHz compared to 20 kHz for the same exposure time). In addition, HPU and thermal treatments (45–60 °C) have synergistic effects on the inactivation of microorganisms [30].

The potential application of HPU technology in the wine industry has also been assessed by Yap et al. [10] and Jiranek et al. [25]. These authors argued that this process could be used for the management of microorganisms at different stages of winemaking. Yap et al. [31] studied the effect of HPU on wood barrels and noticed that HPU removes *Brettanomyces* spp. on the barrel surface and in the stave up to a depth of 4 mm.

Validation tests carried out in 2007 and 2008 by the Australian Wine Research Institute (AWRI) demonstrated that the cleaning and disinfection of barrels (American oak) using HPU technology was more efficient than with steam in conventional conditions of use (1000 psi/6900 kPa at 60 °C for 5 min). In this research, the HPU not only removed all the deposited tartrate, but also inhibited 100% of *Brettanomyces* cells on the surface and up to 4 mm deep in the wood [31]. However, microorganisms can be located deeper in the oak wood (close to 8 mm) and the impact of this treatment on the oak wood structure was not apprehended. Moreover, there were no optimization parameters for the HPU treatment in this study.

Recently, Porter et al. [32] studied the effect of HPU treatment on porous cleaning efficacy in American oak wine barrels using X-ray tomography. It was demonstrated that HPU can significantly remove tartrate deposits from the first two millimeters of oak surfaces, but this was not reproducible at a depth of 2–8 mm. An average of 89% total tartrate volume was removed from the surface layer in the first treatment, but this was further increased to 98% by increasing temperature and time treatment. A highly significant removal of stave surface tartrate crystals was also demonstrated with this cleaning technique at the temperatures studied. Only a few studies considered the oak wood characteristics after HPU treatment, even if oak is a fragile matrix. Oak wood consists of macromolecules such as cellulose, hemicellulose and extractive compounds like ellagitannins, lignin and aromatic precursors. Cellulose and hemicellulose are complex components in the cell wall of wood and make up a large part of oak wood composition (more than 50%). Hemicellulose is formed from covalent bonds with lignin (close to 30% of the total composition) and ester linkage with acetyl units and hydroxycinamic acids.

These bonds are important to ensure the mechanical stability of the structure. These bonds limit the extraction of hemicellulose from the cell wall matrix. The application of ultrasound in extraction and refining processes has drawn increasing attention recently for several applications, and these studies prove that HPU could have a significant impact on cell wall structure [33]. Furthermore, ultrasonic treatment is well established in the processing of plant raw materials, in particular, in extracting low molecular weight substances and pharmaceutically active compounds [34,35]. The authors also consider that HPU could be used to extract cellulose from several plant materials and this aspect is essential for our work. The mechanochemical effect of ultrasound is believed to accelerate the extraction of organic compounds from plant materials due to the disruption of cell walls.

The innovative point of this study is the characterization of the porous material by evaluating the specific surface, oxygen desorption rate, and hydrophobicity of the oak wood after treatment. These parameters are very important as a modification of the oak wood structure could induce several modifications (wettability, oxygen desorption). Indeed, wine might penetrate deeper if the specific surface area and hydrophobicity are changed. Thus, microorganisms could also penetrate further into the depth of the wood and oxygen desorption could be more important. These measurements could ensure that no modifications of the oak wood structure (on the surface and at a depth) are induced by HPU.

The aim of this work was to study the effect of different operating conditions (temperature and time) for HPU treatment on wood properties and sanitation effect. The study investigated the impact of high power ultrasound (HPU) on (i) the specific surface (B.E.T method); (ii) the oxygen desorption kinetics contained in the wood; (iii) the oak wood hydrophobicity (contact angle) and (iv) spoilage microorganism (*Brettanomyces bruxellensis*) removal. All these indications could provide insights into oak wood structure and sanitation possibilities up to 9 mm. Operating parameter optimization was performed in the first part and the microbial stabilization was carried out on with the optimized HPU parameters and the classical barrel treatment (steam) in the second part.

2. Materials and Methods

2.1. Barrel Treatment

2.1.1. High Power Ultrasound (HPU)

The HPU treatment process consisted in filling the barrel with water (heated to 40, 60 or 80 °C by an autonomous system) and then to inserting the sonotrode (Dyogéna, Blanquefort, France) (part that emits the ultrasound waves) into the bung hole (Figure 1), thereby allowing pressurization of the water inside the barrel (0.3 bar). HPU was emitted inside the barrel (frequency 20 kHz, 3.8 kW). The experimental barrel used for the HPU treatment was made in order to place and maintain the experimental staves inside the barrel (Figure 2). The hatch of the experimental barrel was closed for water filling.

HPU experiments were carried out to characterize the influence of operating conditions: temperature and time.

Figure 1. High power ultrasound (HPU) apparatus.

Figure 2. Experimental barrel for HPU treatment.

2.1.2. Aqueous Steam Treatment

The treatment by aqueous steam was carried out with an autonomous boiler Barriclean® (Bouyoud Distribution, Brive-la-gaillarde, France) supplying pressurized hot water during 10 min (1.1 bar, 110 °C) inside the barrel. Steam modalities were only used to compare results for microbiological aspects.

2.2. Lab Experimental Setup and Operating Conditions

The operating conditions investigated for HPU were processing time (4, 6 or 8 min) and temperature (40, 60 or 80 °C). The staves for testing were extracted from French oak barrels with a medium toast used during two years. The study was carried out on 10 types of stave in triplicate, after undergoing various HPU treatments. The control staves were untreated. All the operating conditions are summarized in Table 1. The staves were then cut in different ways to obtain the appropriate sample size for each analysis. All the experiments were carried out in triplicate.

Table 1. Operating condition testing and experimental design.

Treatment Temperature (°C)	Treatment Time (min)	Liquid Used for Treatment	HPU Power (kW)
40	4		
	6		
	8		
60	4	water	3.8
	6		
	8		
80	4		
	6		
	8		
	No treatment		

2.3. Oak Wood Characterization

2.3.1. Determination of Specific Surface

Knowledge of the specific surface of a wood sample (from a stave) is an important parameter to appreciate the exchange surface between the wood and wine. The specific surface refers to the real area of an object, as opposed to its apparent surface area. This is estimated from the amount of nitrogen adsorbed in relation to its pressure at the boiling point of liquid nitrogen and at normal atmospheric pressure.

The SA 3100 BET (Beckman Coulter, Brea, CA, USA) measures the specific surface of granulated samples via gas adsorption using the Brunauer-Emmett-Teller (BET) method. In this so-called discrete method, the data points obtained and the gas pressures are balanced before the reading is recorded. The volume of adsorbed gas retained by the sample is calculated from the pressures recorded at each measuring point.

2.3.2. Oxygen Desorption Staves

Several 500 mL flasks (Schott®) were connected to a vacuum system generating a negative pressure of 0.2 bars. This negative pressure is comparable to that observed in the barrel. In the barrel, the 225 L of wine are in contact with the 2 m^2 of the internal surface of the barrel. In order to simulate the real conditions, the volume surface ratio was respected by introducing an exchange surface of 36.3 cm^2 of wood into each bottle filled with the model solution (12% ethanol v/v, 5 g/L tartaric acid and pH 3.5), made inert beforehand with nitrogen. The external surfaces of the two oak pieces (7.25 cm × 2.5 cm) were covered with silicone gel (Elastosil E43), with the exception of the inner side (toasted side) to reproduce the real conditions in an oak barrel. The inert model solution was changed every 6 days. This methodology was used to avoid the oxygen consumption of the released oak wood polyphenols [36]. Preliminary tests were realized to ensure that no oxygen transfer could occur through the silicone gel. The oxygen desorption kinetics of the wood pieces were monitored every day for 1 month. The oxygen concentration in the liquid flask was detected using a mobile optical fiber coupled with a sensor device. The luminescent system was of the Oxy-trace type (PreSens GmbH, Germany), coupled with a PSt3 type oxygen sensor (detection limit = 15 µg/L, 0–100% oxygen). The spot sensors were placed inside the liquid flask allowing detection of the oxygen concentration.

2.3.3. Contact Angle Measurement

The sessile drop contact angle measurement technique seeks to determine the wettability of the staves studied by characterizing the ease of a liquid drop spreading over a solid surface. We characterized here the hydrophobic/hydrophilic nature of the material. The experimental device used was a Digidrop Contact Angle Meter (GBX Scientific Instruments, Ireland). A 20 µL water droplet was placed on the stave and the contact angle was measured after stabilization for around 200 s. The contact angle is defined as the angle between the solid surface and a tangent, drawn on the drop surface, passing through the atmosphere–liquid–solid triple point (Figure 3) [37]. An image system analysis integrates the angle as a function of time and makes it possible to determine whether the drop enters the material.

For each modality studied, the contact angles were calculated from measurements on three pieces of stave from three different parts of each stave. The drop was systematically in a place where there were no apparent residues of tartrate, to avoid any experimental artefacts. For each experimental modality, the average contact angle was thus obtained from 9 individual measurements on three different samples.

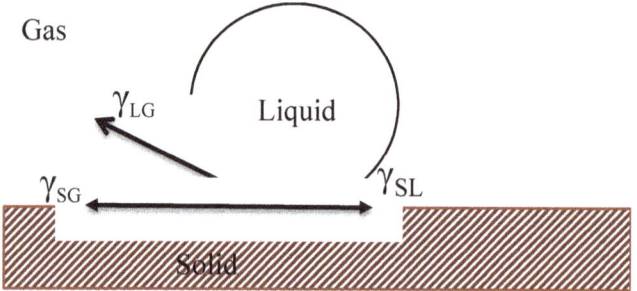

Figure 3. Schematic example of contact angle measurement.

2.4. Impact on Microorganisms

The impact of HPU on spoilage microorganisms, especially *Brettanomyces bruxellensis*, was investigated according to their depth in the wood.

The staves of French oak wood barrels of one or two years (medium toast) were incubated in a liquid culture of YPG (Yeast extract 10 g/L; bactopeptone 10 g/L; glucose 20 g/L; adjusted to pH 5), supplemented with antibiotics in order to limit the growth of bacteria, molds and yeast of the *Saccharomyces* genus (0.1 g/L chloramphenicol; 0.15 g/L biphenyl; 0.5 g/L cycloheximide), and containing *B. bruxellensis* L0539 (available through the "Centre de Ressources Biologiques Œnologiques" of Bordeaux University (CRBO)) in mid-exponential phase during 4 days at room temperature. The population was determined before and after treatment by drilling staves to different depths (0–2 mm; 2–5 mm; 5–9 mm) with a small drill bit and permits recovering 0.2 g of wood.

The wood samples recovered at different depths were incubated in 2 mL of sterile saline solution (9 g/L sodium chloride) during 48 h at room temperature under agitation. Serial dilutions of these samples were plated on solid YPG (Yeast extract 10 g/L; bactopeptone 10 g/L; glucose 20 g/L; agar 20 g/L; adjusted to pH 5) supplemented with antibiotics (0.1 g/L chloramphenicol; 0.15 g/L biphenyl; 0.5 g/L cycloheximide). Yeast populations in the primary dilution were monitored by fluorescence microscopy.

Colonies were counted after 7 days of incubation at 30 °C. All assays were performed in triplicate.

2.5. Statistical Analyses

Statistical data were analyzed using the Kruskal–Wallis non-parametric test (RStudio software, v1.0.143, RStudio Inc., Boston, USA, http://www.rstudio.com/) to identify the means that were significantly different. The statistically significant level was 5% ($p < 0.05$).

3. Results and Discussion

3.1. Specific Surface Area

The objective of this measurement is to characterize the specific surface area, which could have an impact on the liquid/solid exchange surface. The specific surface area of a wood sample (from a stave) could be defined as the total surface area of a material per unit of mass and has a particular importance for adsorption phenomena.

The specific surface area measured by the BET method for the different test modalities tested is presented in Figure 4. The specific surface area increases regardless of the HPU treatment modalities compared to the control (from 900% to 1400%). There is no effect of the time treatment on the specific surface contrarily to temperature. Indeed, at high temperature (80 °C), we notice that the specific surface area was around 2 m²/g, which is significantly lower than 40 °C and 60 °C. These low values compared to other temperatures should indicate a wood modification of HPU treatment above 60 °C. This deterioration of wood integrity could involve a degradation of lignin, cellulose and hemicellulose [38]. Moreover the increase in the specific area could also indicate that tartrate is effectively removed from the wood structure during HPU treatment, especially for 40 °C and 60 °C, while remaining present in the untreated controlled staves. Figure 5 illustrates an example of tartrate removal from HPU treatment at 60 °C. However, the surface tartrate is more efficiently removed at 60 °C and 80 °C, instead of 40 °C. These results corroborate those of Porter et al. [32], which showed 98% of total tartrate volume removed with HPU treatment (4 KW, 12 min at 40–60 °C).

Ultrasound treatment clearly modifies the physiochemical structure of wood because we notice significant differences between all the HPU modalities and the control. These considerations were also observed by He et al. [38]. These authors considered that HPU could decrease the alkali metals in the resulting material, and significantly increase its crystallinity. It has been observed in the case of eucalyptus, and our results indicate that the same trend is obtained for oak wood. The authors considered that a rupture appears between the methyl/methylene groups in cellulose and contributes to removing cellulose, hemicellulose and lignin.

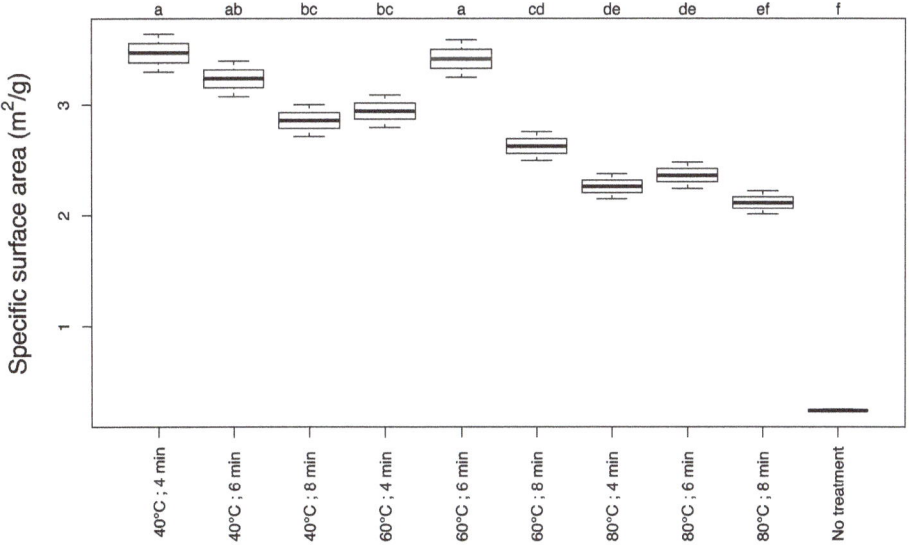

Figure 4. Specific surface of wood treated by HPU.

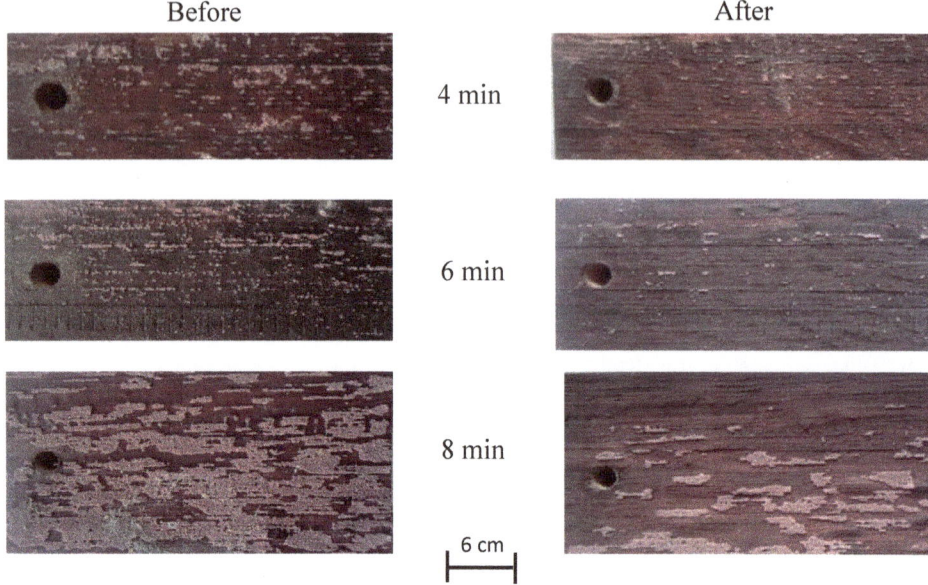

Figure 5. Examples of tartrate removal before and after HPU treatment (60 °C).

HPU also apparently increases the exposure of the material to the treatment solution and enhances its accessibility, as well as breaking down pits, which could generate collapses and micro channels, and removing attachments on the wood tissue. This aspect will have an impact on oxygen desorption and should be considered to validate our results. Finally, further investigation should be made on the specific surface deeper in the wood stave.

3.2. Oxygen Desorption of Staves

The influence of the HPU treatment on the quantity of oxygen desorbed and its kinetics was investigated with a specific vacuum system. The O_2 desorption monitoring was carried out over 26 days for each treatment considered in triplicate. The results for HPU treatment at 60 °C are presented in Figure 6.

We noticed that untreated staves desorbed less oxygen over the first 6 days. We could consider in this case that tartrate is still present in the oak wood vessels and thus could limit the oxygen desorption kinetic. For HPU treatment, in the first 6 days we noticed that O_2 desorption is higher for 8 min than for 6 min at 60 °C. Then, over the next 6 days, we observed that desorption of HPU 6 min was slightly higher than HPU 8 min, with a variation close to 0.5 mg/L. We can see in Table 2 that there was no difference in total oxygen transfer over 26 days for HPU treatment between 6 and 8 min. On the other hand, the total oxygen concentration desorbed was 5.57 ± 1.25 mg/L for untreated stave, which was significantly lower than HPU. In comparison to results obtained by Qiu et al. [39], the O_2 desorption rate for wood barrel during the first month was lower in this case (close to 10 mg/L for a new oak wood [36]). Considering other HPU temperatures (40 °C and 80 °C), the same trend could be observed (results not shown).

Our results indicate that oxygen desorption is highly impacted by HPU treatment. In the case of untreated oak wood, the oxygen desorption rate was two times lower, as was kinetics, especially for the first week. After HPU treatment, the sum was close to 10 mg/L indicating a significant variation. These values are still similar to an unused oak wood, which indicates that the oak structure is not impacted. The hypothesis of the appearance of micro channels was not verified here for the HPU treatment at 60 °C. Oak wood ultrastructure seems to be conserved.

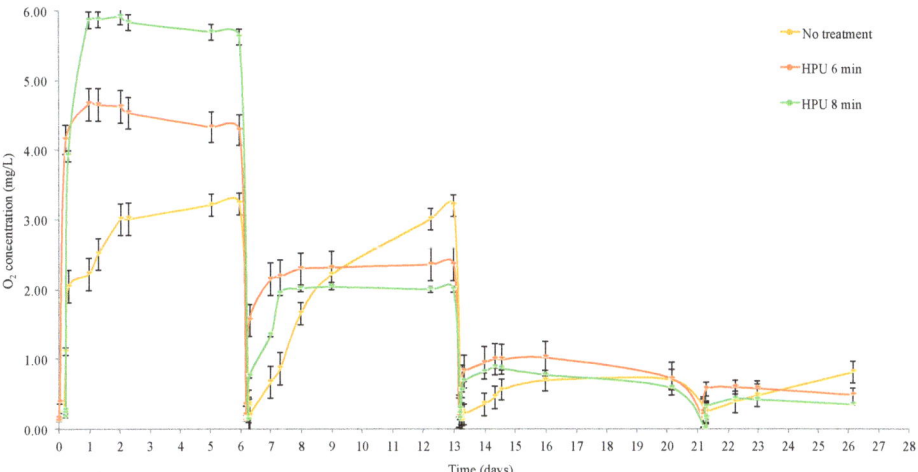

Figure 6. Influence of HPU treatment on O_2 desorption.

Table 2. Total oxygen desorption for different type of treatment over 26 days. The total oxygen concentration is the sum of each mean oxygen value during each 6-day period previously shown in Figure 6.

Type of Treatment	Total Oxygen Concentration (mg/L)
No treatment	5.57 ± 1.05 [a]
HPU 60 °C 6 min	8.11 ± 1.15 [b]
HPU 60 °C 8 min	8.63 ± 0.43 [b]

Subscript letter refers to the significant differences between the types of treatment ($p = 0.05$).

3.3. Contact Angle

The purpose of this analysis was to characterize the hydrophobic/hydrophilic nature of the contact surface. The contact angle could indicate the modifications that may possibly occur at the structural level of the staves after HPU treatment. By definition, the smaller the contact angle, the higher the hydrophobicity, which could have an impact on the absorption rate.

The contact angle measurements for HPU treatments are presented in Figure 7. We could consider that the duration of treatment does not seem to have a decisive influence on the hydrophobic nature of the wood surface. However, we noticed a trend in the HPU effect with the global increase in contact angle values. In the other types of treatment, no significant difference was observed compared to the control, although we did notice an upward trend in the contact angle. These results suggest that wood samples are more hydrophobic and significant differences are observed for some cases (80 °C and 40 °C/6 min). These differences are possibly related to lignin removal, the presence of hemicelluloses or other carbohydrate material and extractives at the fiber surface. These results are similar to those obtained by [33]. In our case, our results suggest that, HPU treatments and, more especially, high temperature (80 °C), could induce some modification because the contact angle is significantly different (higher than 50° for each case). The results obtained are in good agreement with those observed for the main wood components, reported by Young [40]. The authors proved the wettability of wood pulp fibers where hardwood lignin (kraft) has a contact angle of 60° and cellulose 33–34°. The authors consider that surfaces rich in lignin and extractives have higher contact angles, and the values obtained in this case were close to 60°. These values therefore indicate a greater proportion of extractives and lignin in the surface of oak wood, which is a more hydrophobic surface. The potential of ultrasound to extract polysaccharide components has been widely studied in different plants and plant tissues and this phenomenon is confirmed in our study. Ultrasounds are known to be a powerful tool for accelerating polysaccharide extraction. In our case, the extraction seems to be effective because the hydrophobic characteristics are increased in the case of HPU treatment at high temperature. This could indicate that polysaccharide content is also increased on the surface and their desorption will be higher. This parameter is essential because it will impact the oak wood wettability. In our case, the use of HPU generally led to an increase in the hydrophobicity of the wood (from 125% to 350%). Even if the link between wettability and O_2 desorption kinetics exists from a theoretical point of view, it is difficult to extrapolate in our case and additional experiments should be conducted to validate this hypothesis.

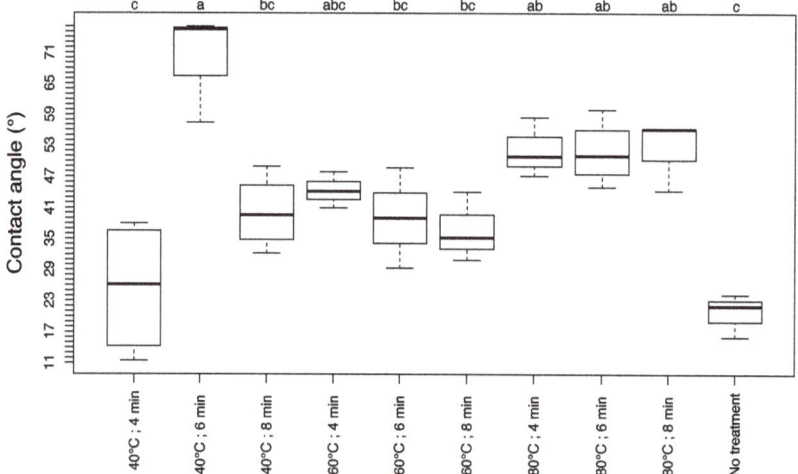

Figure 7. Influence of HPU treatments on wood hydrophobicity by contact angle measurements.

3.4. Sanitation Effect on Spoilage Microorganism B. Bruxellensis

Sanitation effects of two types of treatment (HPU 6 min/60 °C and steam 10 min/110 °C) were investigated for the removal of the spoilage microorganism *B. bruxellensis*. Staves inoculated with *B. bruxellensis* were treated with HPU and different depth samples were recovered in order to estimate the effectiveness of the in-depth treatment as we can see in Table 3.

If we focus on HPU treatment (6 min/60 °C), we notice that the post-treatment population of *B. bruxellensis* is lower than the detection limit (1 log CFU/g) to a depth close to 9 mm. This length represents the maximum depth reached by the wine by passive diffusion in the wood during aging and therefore corresponds to the maximum depth reached by *Brettanomyces* yeast. This efficiency is probably due to the synergistic effect between the HPU treatment, especially cavitation bubbles, and the thermal effect of the water at 60 °C brought into contact with yeast nested in the wood. On the other hand, we consider that steam treatment is less efficient because the post-treatment population of *B. bruxellensis* is unchanged from 2 mm to 9 mm. The depth efficiency for steam treatment is 2 mm with a population lower than the detection limit. We could consider that these results are due to the thermal inertia of the wood. During steam treatment, the first millimeters of the wood reach a temperature close to 100 °C, allowing elimination of *Brettanomyces* yeasts, whereas the temperature only reaches 45 °C at 5 mm. We can also see that there is no difference in the treatment response (HPU or steam) between staves of 1 or 2 years. Yap et al. [31] have studied the effect of HPU on wood barrels and noticed that HPU removed *Brettanomyces* spp. on the barrel surface and in the stave up to

4 mm, but they did not investigate to further depths. In comparison to another innovative treatment, González-Arenzana et al. [14] studied the microwaves capacity to sterilize French oak wood barrels and they were able to remove 35% of the population for *Brettanomyces* spp., 36% of total yeast, 90% of lactic acid bacteria and 100% of acetic bacteria up to a depth 8 mm [14]. In our case with optimized operating parameters, the HPU sanitation effect reaches a depth of 9 mm in the wood, which is also the depth to which the wine and therefore *Brettanomyces* yeast can penetrate.

According to these results, the barrels contaminated by *B. bruxellensis* can be reused if they are treated with HPU (3.8 kW, 6 min at 60 °C), unlike steam treatment that leaves viable *B. bruxellensis* cells in the depth of the wood, which can quickly become sources of recontamination. The HPU treatment is expected to target cells that are located deep within the pores of the staves, which would otherwise be untreated by classical barrel treatments.

Table 3. *B. bruxellensis* population before and after HPU or steam treatment on staves of one and two years at different sampling depth.

Stave Age (year)	Type of Treatment	Sampling Depth (mm)	*B. bruxellensis* Population before Treatment (log CFU/g)	*B. bruxellensis* Population Post Treatment (log CFU/g)
1	HPU 6 min 60 °C	0–2	7.73 ± 0.02	<DL
		2–5	5.89 ± 0.04	<DL
		5–9	4.23 ± 0.01	<DL
	Steam 10 min 110 °C	0–2	7.79 ± 0.05	<DL
		2–5	5.71 ± 0.03	4.92 ± 0.04
		5–9	4.63 ± 0.02	4.59 ± 0.03
2	HPU 6 min 60 °C	0–2	8.11 ± 0.04	<DL
		2–5	6.08 ± 0.02	<DL
		5–9	5.61 ± 0.01	<DL
	Steam 10 min 110 °C	0–2	7.91 ± 0.05	<DL
		2–5	6.82 ± 0.03	5.96 ± 0.02
		5–9	5.71 ± 0.02	5.63 ± 0.03

<DL: Detection limit (1 log CFU/g).

4. Conclusions

This study has shown that the combined effect of HPU and heat treatment may have an impact on wood sanitation, the wettability of wood, its specific surface and oxygen transfer kinetics. The operating parameters used during the HPU treatment are essential.

Specific surface measurement seems to be a relevant method for determining the tartrate removal efficiency of wood. This method has good repeatability and is not influenced by the heterogeneity of the wood surface. Concerning the hydrophobicity of the wood, we have shown that HPU could increase the contact angle, especially at high temperatures (80 °C). Thus, the HPU treatment enables the initial oxygen transfer capacity of the wood to be partially recovered particularly because of the surface tartrate removal. Nevertheless, we could go further in the experiments by extending to different wood origin and types of barrels (age, toast, wine, wood grain) or by investigating the ultrastructure of the wood.

Finally, the sanitation effect of HPU was investigated and permits removal of all viable *B. bruxellensis* cells up to a depth 9 mm with processing parameters set at 60 °C/6 min with 3.8 kW. These parameters (60 °C; 6 min) are the most efficient in regards to all of these issues.

Author Contributions: P.R. and R.G. conceived the experimental design; F.M. performed the contact angle, specific surface and oxygen desorption experiment; P.R. carried out the HPU and steam treatment, as well as the microbiology test; P.R. and R.G. supervised the experiments; P.R. and M.B. processed the results; M.B. wrote the paper: R.G., P.R. and F.M. reviewed the manuscript.

Funding: This research has been financed by the Region Aquitaine.

Acknowledgments: The authors would like to thank Hancock Hutton Langues Services for the English corrections. We would like to thank Dyogena and the Region Aquitaine for the financial support.

Conflicts of Interest: The authors declare no conflict of interest.

References

1. Cerdán, T.G.; Ancín-Azpilicueta, C. Effect of oak barrel type on the volatile composition of wine: Storage time optimization. *LTW Food Sci. Technol.* **2006**, *39*, 199–205. [CrossRef]

2. Ortega-Heras, M.; González-Sanjosé, M.L.; González-Huerta, C. Consideration of the influence of aging process, type of wine and oenological classic parameters on the levels of wood volatile compounds present in red wines. *Food Chem.* **2007**, *103*, 1434–1448. [CrossRef]

3. Chira, K.; Teissedre, P.-L. Chemical and sensory evaluation of wine matured in oak barrel: Effect of oak species involved and toasting process. *Eur. Food Res. Technol.* **2014**, *240*, 533–547. [CrossRef]

4. González-Centeno, M.R.; Chira, K.; Teissedre, P.-L. Comparison between Malolactic Fermentation Container and Barrel Toasting Effects on Phenolic, Volatile, and Sensory Profiles of Red Wines. *J. Agric. Food Chem.* **2017**, *65*, 3320–3329. [CrossRef] [PubMed]

5. Malfeito-Ferreira, M.; Laureano, P.; Barata, A.; Antuono, I.D.; Stender, H.; Loureiro, V. Effect of different barrique sanitation procedures on yeasts isolated from the inner layers of Wood. In Proceedings of the ASEV 55th Annual Meeting, San Diego, CA, USA, 29–30 June 2004.

6. Cibrario, A. Diversité Génétique et Phénotypique de L'espèce Brettanomyces bruxellensis: Influence sur son Potentiel D'altération des Vins Rouges. Ph.D. Thesis, Université de Bordeaux, Bordeaux, France, 2017.

7. Breniaux, M.; Dutilh, L.; Petrel, M.; Gontier, E.; Campbell-Sills, H.; Deleris-Bou, M.; Krieger, S.; Teissedre, P.L.; Jourdes, M.; Reguant, C.; et al. Adaptation of two groups of *Oenococcus oeni* strains to red and white wines: The role of acidity and phenolic compounds. *J. Appl. Microbiol.* **2018**, *125*, 1117–1127. [CrossRef] [PubMed]

8. Chatonnet, P.; Dubourdieu, D.; Boidron, J.N.; Pons, M. The origin of ethylphenols in wines. *J. Sci. Food Agric.* **1992**, *60*, 165–178. [CrossRef]

9. Coulon, J.; Perello, M.C.; Funel, A.L.; de Revel, G.; Renouf, V. *Brettanomyces bruxellensis* evolution and volatile phenols production in red wines during storage in bottles. *J. Appl. Microbiol.* **2010**, *108*, 1450–1458. [CrossRef]

10. Yap, A.; Jiranek, V.; Grbin, P.; Barnes, M.; Bates, D. Studies on the application of high-power ultrasonics for barrel and plank cleaning and disinfection. *Aust. N. Z. Wine Ind. J.* **2007**, *22*, 96–104.

11. Conterno, L.; Joseph, C.; Arvik, T.; Henick-Kling, T.; Bisson, L.F. Genetic and physiological characterization of *Brettanomyces bruxellensis* strains isolated from wines. *Am. J. Enol. Viticult.* **2006**, *57*, 139–147.

12. Curtin, C.D.; Bellon, J.R.; Henschke, P.A.; Godden, P.W.; de Barros Lopes, M.A. Genetic diversity of *Dekkera bruxellensis* yeasts isolated from Australian wineries. *FEMS Yeast Res.* **2007**, *7*, 471–481. [CrossRef]

13. Guzzon, R.; Widmann, G.; Malacarne, M.; Nardin, T.; Nicolini, G.; Larcher, R. Survey of the yeast population inside wine barrels and the effects of certain techniques in preventing microbiological spoilage. *Eur. Food Res. Technol.* **2011**, *233*, 285–291. [CrossRef]

14. González-Arenzana, L.; Santamaría, P.; López, R.; Garijo, P.; Gutiérrez, A.R.; Garde-Cerdán, T.; López-Alfaro, I. Microwave technology as a new tool to improve microbiological control of oak barrels: A preliminary study. *Food Control.* **2013**, *30*, 536–539. [CrossRef]

15. Poupault, P.; Richard, R. Bio-Adhésion des Levures du Genre Brettanomyces et Conséquences sur L'hygiène de la Barrique. *Matevi-France.com* **2006**, *76*, 1–5. Available online: http://www.matevi-france.com/fileadmin/user_upload/fichiers_matevi/Autres_materiels_pdf/Bio-adhesion_microbienne_et_hygiene_des_barriques_Matevi_2016.pdf (accessed on 5 December 2018).

16. Mawson, R.; Knoerzer, K. A brief history of the application of ultrasonics in food processing. In Proceedings of the 19th International Congress on Acoustics, Madrid, Spain, 2–7 September 2007.

17. Patist, A.; Bates, D. Ultrasonic innovations in the food industry: From the laboratory to commercial production. *Innov. Food Sci. Emerg. Technol.* **2008**, *9*, 147–154. [CrossRef]

18. McClements, D.J. Advances in the application of ultrasound in food analysis and processing. *Trends Food Sci. Technol.* **1995**, *6*, 293–299. [CrossRef]

19. Leighton, T. *Ultrasound in Food Processing*; Springer Science & Business Media: Berlin, Germany, 1998.

20. Villamiel, M.; de Jong, P. Influence of high-intensity ultrasound and heat treatment in continuous flow on fat, proteins, and native enzymes of milk. *J. Agric. Food Chem.* **2000**, *48*, 472–478. [CrossRef]

21. Maisonhaute, E.; Prado, C.; White, P.; Compton, R.G. Surface acoustic cavitation understood via nanosecond electrochemistry. Part III: Shear stress in ultrasonic cleaning. *Ultrason. Sonochem.* **2002**, *9*, 297–303. [CrossRef]
22. Krefting, D.; Mettin, R.; Lauterborn, W. High-speed observation of acoustic cavitation erosion in multibubble systems. *Ultrason. Sonochem.* **2004**, *11*, 119–123. [CrossRef]
23. Leighton, T. What is ultrasound? *Prog. Biophys. Mol. Biol.* **2007**, *93*, 3–83. [CrossRef]
24. Piyasena, P.; Mohareb, E.; McKellar, R. Inactivation of microbes using ultrasound: A review. *Int. J. Food Microbiol.* **2003**, *87*, 207–216. [CrossRef]
25. Jiranek, V.; Grbin, P.; Yap, A.; Barnes, M.; Bates, D. High power ultrasonics as a novel tool offering new opportunities for managing wine microbiology. *Biotechnol. Lett.* **2008**, *30*, 1–6. [CrossRef] [PubMed]
26. Guerrero, S.; López-Malo, A.; Alzamora, S.M. Effect of ultrasound on the survival of *Saccharomyces cerevisiae*: Influence of temperature, pH and amplitude. *Innov. Food Sci. Emerg. Technol.* **2001**, *2*, 31–39. [CrossRef]
27. Furuta, M.; Yamaguchi, M.; Tsukamoto, T.; Yim, B.; Stavarache, C.E.; Hashiba, K.; Maeda, Y. Inactivation of Escherichia coli by ultrasonic irradiation. *Ultrason. Sonochem.* **2004**, *11*, 57–60. [CrossRef]
28. Tsukamoto, I.; Yim, B.; Stavarache, C.E.; Furuta, M.; Hashiba, K.; Maeda, Y. Inactivation of *Saccharomyces cerevisiae* by ultrasonic irradiation. *Ultrason. Sonochem.* **2004**, *11*, 61–65. [CrossRef]
29. Borthwick, K.A.J.; Coakley, W.T.; McDonnell, M.B.; Nowotny, H.; Benes, E.; Gröschl, M. Development of a novel compact sonicator for cell disruption. *J. Microbiol. Methods* **2005**, *60*, 207–216. [CrossRef]
30. López-Malo, A.; Palou, E.; Jiménez-Fernández, M.; Alzamora, S.M.; Guerrero, S. Multifactorial fungal inactivation combining thermosonication and antimicrobials. *J. Food Eng.* **2005**, *67*, 87–93. [CrossRef]
31. Yap, A.; Schmid, F.; Jiranek, V.; Grbin, P.; Bates, D. Inactivation of *Brettanomyces/Dekkera* in wine barrels by high power ultrasound. *Aust. N. Z. Wine Ind. J.* **2008**, *23*, 32–40.
32. Porter, G.W.; Lewis, A.; Barnes, M.; Williams, R. Evaluation of high power ultrasound porous cleaning efficacy in American oak wine barrels using X-ray tomography. *Innov. Food Sci. Emerg. Technol.* **2011**, *12*, 509–514. [CrossRef]
33. Hromádková, Z.; Ebringerová, A. Ultrasonic extraction of plant materials—investigation of hemicellulose release from buckwheat hulls. *Ultrason. Sonochem.* **2003**, *10*, 127–133. [CrossRef]
34. Mason, T.; Paniwnyk, L.; Lorimer, J.P. The uses of ultrasound in food technology. *Ultrason. Sonochem.* **1996**, *3*, S253–S260. [CrossRef]
35. Sališová, M.; Toma, Š.; Mason, T.J. Comparison of conventional and ultrasonically assisted extractions of pharmaceutically active compounds from Salvia officinalis. *Ultrason. Sonochem.* **1997**, *4*, 131–134. [CrossRef]
36. Qiu, Y. Phénomènes de Transfert D'oxygène à Travers la Barrique. Ph.D. Thesis, Université de Bordeaux, Bordeaux, France, 2015.
37. Gindl, M.; Sinn, G.; Gindl, W.; Reiterer, A.; Tschegg, S. A comparison of different methods to calculate the surface free energy of wood using contact angle measurements. *Colloids Surf. A: Physicochem. Eng. Aspects* **2001**, *181*, 279–287. [CrossRef]
38. He, Z.; Wang, Z.; Zhao, Z.; Yi, S.; Mu, J.; Wang, X. Influence of ultrasound pretreatment on wood physiochemical structure. *Ultrason. Sonochem.* **2017**, *34*, 136–141. [CrossRef] [PubMed]
39. Qiu, Y.; Lacampagne, S.; Mirabel, M.; Peuchot, M.M.; Ghidossi, R. Oxygen desorption and oxygen transfer through oak staves and oak stave gaps: An innovative permeameter. *OENO ONE* **2018**, *52*, 1–14. [CrossRef]
40. Young, R.A. Wettability of Wood Pulp Fibers: Applicability of Methodology. *Wood Fiber Sci.* **2007**, *8*, 120–128.

Article

Oxygen Consumption by Red Wines under Different Micro-Oxygenation Strategies and *Q. Pyrenaica* Chips. Effects on Color and Phenolic Characteristics

Rosario Sánchez-Gómez ⓘ, Ignacio Nevares ⓘ, Ana María Martínez-Gil ⓘ and Maria del Alamo-Sanza * ⓘ

Grupo UVaMOX, E.T.S. Ingenierías Agrarias, Universidad de Valladolid, Avda. Madrid 50, 34004 Palencia, Spain; rosario.sanchez@uva.es (R.S.-G.); ignacio.nevares@uva.es (I.N.); anamaria.martinez.gil@uva.es (A.M.M.-G.)
* Correspondence: maria.alamo.sanza@uva.es

Received: 9 August 2018; Accepted: 5 September 2018; Published: 6 September 2018

Abstract: The use of alternative oak products (AOP) for wine aging is a common practice in which micro-oxygenation (MOX) is a key factor to obtain a final wine that is more stable over time and with similar characteristics as barrel-aged wines. Therefore, the oxygen dosage added must be that which the wine is able to consume to develop correctly. Oxygen consumption by red wine determines its properties, so it is essential that micro-oxygenation be managed properly. This paper shows the results from the study of the influence on red wine of two different MOX strategies: floating oxygen dosage (with dissolved oxygen setpoint of 50 µg/L) and fixed oxygen dosage (3 mL/L·month). The results indicated that the wines consumed all the oxygen provided: those from fixed MOX received between 3 and 3.5 times more oxygen than the floating MOX strategy, the oxygen contribution from the air entrapped in the wood being more significant in the latter. Wines aged with wood and MOX showed the same color and phenolic evolution as those aged in barrels, demonstrating the importance of MOX management. Despite the differences in the oxygen consumed, it was not possible to differentiate wines from the different MOX strategies at the end of the aging period in contact with wood.

Keywords: aging; chips; dissolved oxygen; floating and fixed micro-oxygenation; *Quercus pyrenaica*; red wine

1. Introduction

Oxygen has a fundamental role in wine technology [1]. It plays an important role in the different processes that take place during wine-making and aging [2–6]. Oak barrel aging is traditionally used in wine-making to produce high quality wines, since the contact between wine and oxygen in the oak barrels influences its composition.

In order to shorten time and reduce costs [7–10], the use of alternative oak products (AOP) is widespread. Although its combination with the micro-oxygenation (MOX, small oxygen dosage) technique has scarcely been used until now [11,12], it is known that its results essential to reproducing the behavior, and hence the benefits, of the barrel. Since micro-oxygenation is the controlled introduction of oxygen into wine, the dosage and duration of oxygen addition are the critical points in MOX treatment, and positive effects can be obtained when the treatment is applied correctly. This can be acquired by specifying the oxygen management for each kind of wine, alternative oak product (chips, cubes and staves, among others) and also for the botanical origin of each wood [13]. Active micro-oxygenation could be accomplished in two ways: (i) by means of continually adding small fixed dosages of oxygen, known as fixed MOX dosage; and (ii) by means of an adaptive dosage at the level of dissolved oxygen (DO) present in the wine, known as floating MOX dosage. The latter strategy consists of an adaptive oxygen dosage to achieve the desired amount of DO in the wine (setpoint). This content can satisfy the demand throughout the aging process, and must be maintained throughout it to ensure the best integration of wood and wine. Thus, the dosage can be regulated by comparing the reading of each DO measurement with the reference of the DO level.

Recently, it has been demonstrated that, when oak chips were flooded with wine, they would provide 0.135 mg of oxygen per gram of oak chips [14]. In wine aging processes, the oxygen contained in alternative products (oak chips, staves, cubes, etc.), added to wine, has to be estimated correctly. The wine needs to count on the oxygen necessary to evolve appropriately during these processes of aging with wood products, so the dosage of oxygen provided by the wood itself needs to be added to that added by active or passive MOX [15].

Quercus pyrenaica wood's effectiveness in wine aging has meant that it is considered highly advisable as a source of barrels [16]. However, its forest management does not allow it to be supplied to the barrel manufacture industry, but as a source for obtaining alternative products [11,17–21].

The main goal of this work was to study the influence on red wine of two different strategies of MOX/floating oxygen dosage (with a dissolved oxygen setpoint of 50 μg/L) and fixed oxygen dosage (3 mL/L·month), together with the effect over time of adding chips of *Q. pyrenaica* oak wood during aging: all at the beginning, or fractionations at two different times during the process.

2. Materials and Methods

2.1. Wood Samples

Oak heartwood in the form of chips (1 cm × 0.5 cm, approximately) from *Q. pyrenaica* trees, grown in Salamanca (Spain) and provided by CESEFOR (Soria, Spain), were used after natural seasoning in climatic conditions. The wood was then toasted in an industrial-scale convection oven located in the experimental cellar of the University of Valladolid (Palencia, Spain), with supports specially adapted to special oven trays for chips (BINDER APT-COM V 1.0., New York, NY, USA) at 190 °C for 10 min.

2.2. Wine

A young red wine made from a red single-variety grape (cv. Tinta del País) belonging to the Spanish appellation of origin Ribera del Duero, and produced on an industrial scale in 2008, was treated using different MOX aging systems for a period of 4 months. The chemical parameters of the wine before aging were: total acidity 6.1 g/L (expressed as tartaric acid), volatile acidity 0.69 g/L (expressed as acetic acid), sugars 1.33 g/L, degree of alcohol 14.59%, color intensity 21, and total polyphenol 2.2 g/L (expressed as gallic acid). These parameters were evaluated before the wine was transferred into tanks, and also during aging, in accordance with International Organization of Vine and Wine (OIV) methods (OIV, 1990).

The wines were transferred into the tanks and samples were taken from each after 20, 48, 76, 97 and 111 days' aging. After 111 days, the wines were taken out of the tanks and bottled separately. Samples were taken periodically from each aging system.

2.3. Wood and Micro-Oxygenation (MOX) Strategies

Oak chips were added to the tanks at two different moments: in half of them, all the wood was added at the beginning of the aging process, whereas in the other half of the tanks, the wood was added twice, half at the beginning and the other half 48 days after the beginning of the experiment, coinciding with a wine sampling.

The quantity of oak chips added was calculated using the surface/volume relation of 225-L barrels in order to determine the quantity of oak chips necessary to reproduce the same relation in 225-L stainless steel tanks [22]. The oak chip dosage was determined by their weight distributed over a known surface: 1250 g of oak chips were used for each tank. Two different strategies were applied: A—all oak chips at the beginning of the experiment; and B—half of the oak chips at the beginning and the other half 48 days later. All the tanks with oak chips were micro-oxygenated using an Eco2 device (Oenodev, Maumusson-Laguian, France) and ceramic diffusers. The MOX dosage was: A—floating MOX strategy (FMOX), in which the setpoint was set at 50 µg/L; B—fixed MOX strategy, with 3 mL/L·month. Additionally, the quantity of air in the oak chips' interior was taken into account, since that is an additional oxygen contribution, which can be estimated at 0.135 mg oxygen per gram of oak chips [14]. The quantity of oxygen contributed during initial filling of the tanks was added and established as 1 mg/L.

The wine was stored in stainless steel 225 L tanks with oak chips and MOX (Figure 1). Every aging system was replicated, thus requiring eight stainless steel tanks as follows: *tanks 1* and *2* with A wood strategy and A MOX dosage [13], *tanks 3* and *4* with B wood strategy and A MOX dosage, *tanks 5* and *6* with A wood strategy and B MOX dosage, and *tanks 7* and *8* with B wood strategy and B MOX dosage. The wines were matured in the same aging room in the experimental cellar of the University of Valladolid (Palencia, Spain), where humidity and temperature conditions were controlled at 65–75% and 15–16 °C, ensuring MOX in the best way over the aging period [3,23].

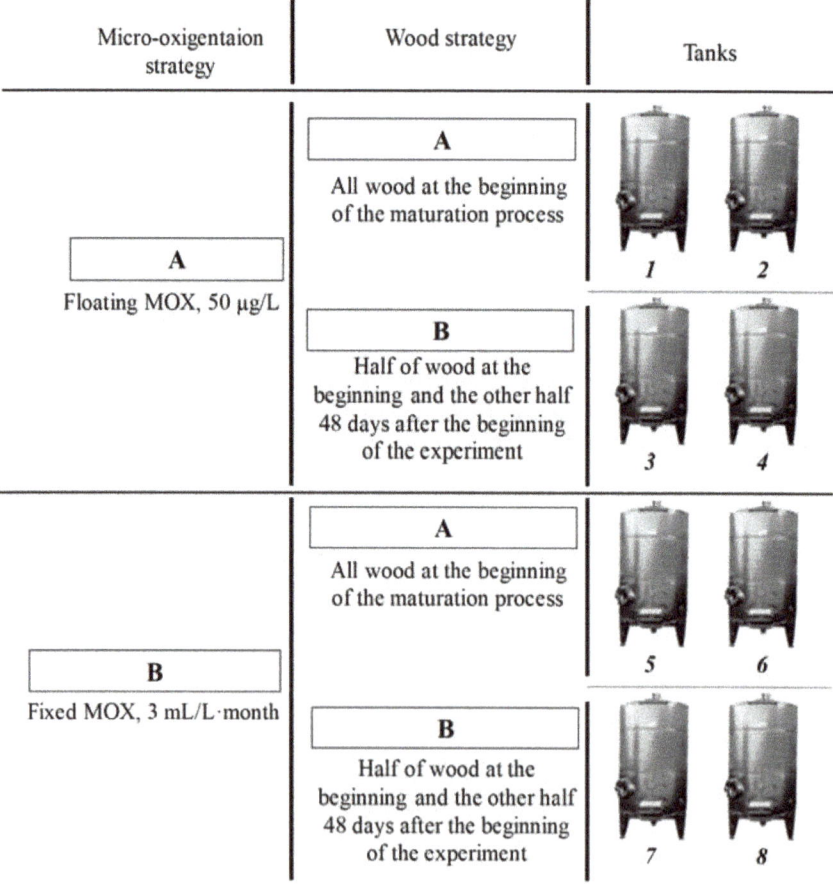

Figure 1. Wine micro-oxygenation (MOX) and wood strategies scheme.

The dosages were checked daily with each DO measurement and, in the case of the floating MOX strategy, adjusted, increased or decreased according to the DO level of each reading, always searching for the DO setpoint.

2.4. Oxygen Determination

The DO measurement system used was capable of measuring oxygen concentrations at µg/L (ppb) level. An Electrochemical system was selected, model 3650/111 Micrologger O_2 (Orbisphere Laboratories, Geneva, Switzerland) and equipped with a sensor measuring from 0.1 µg/L (ppb) to 80 mg/L (ppm) with an accuracy ± 0.001 mg/L, and a detection limit from 0.1 µg/L using the most sensitive membrane. It was fitted in a stainless-steel flow chamber. The conditions of the micro-oxygenated wines were not altered because a non-intrusive pumping system, based on a small peristaltic reversible turn pump equipped with a Tygon® tube, was used. The flow rate was 10 mL/min in order to avoid the influence of any oxygen consumption by the probe or oxygen diffusion when low oxygen concentrations of samples needed to be measured at very low flow rates (≤0.1 mL/min). This enabled the use of a tangential flux, essential for correctly measuring the DO [24]. The whole system was argon-inerted and equipped with quick-connectors that linked all the tanks with non-permeable flexible tubing in order to avoid any interference. Wine samplings were collected at mid-height from each tank. Wine was force-returned to the tanks with argon, in order to guarantee no oxygenation of the system at any time. This procedure was previously tested to verify the absence of oxygen permeability through the walls of the tubing or the fittings [23].

The electrochemical equipment was calibrated in air as a valid option before every reading; in addition, zero calibration was performed in electrochemical systems between every tank reading when the installation was purged with argon to return the wine to the tank. The sensors were chemically reconditioned and the membranes replaced if the zero calibration was above 2 µg/L [25]. A detailed calibration in air-saturated water at a constant temperature, as described in other studies, was carried out monthly [26].

2.5. Consumed Oxygen Determination

The oxygen consumed by the wine aged in stainless steel tanks was calculated every reading day by the difference between the oxygen dosage and the remaining dissolved oxygen at every moment.

2.6. Wine Analysis

2.6.1. Phenolic, Anthocyanin, and Tannin Global Parameters Determination

Phenolic compounds, such as total phenols (PT, as mg/L of gallic acid), were determined by the method of Folin-Ciocalteu [27], and low polymerized phenols (LPP, as mg/L of gallic acid) were determined by Masquelier et al. [28]. High polymerized phenols (HPP, as mg/L of gallic acid) were calculated by the difference between PT and LPP. Total anthocyanins (ACY, as mg/L of malvidin-3-*O*-glucoside) were analyzed by means of color changes according to the pH of the medium [29], tannins (TAN, as g/L of cyanidin chloride) using the Ribéreau-Gayón and Stonestreet method [30] and, finally, catechins (CAT, as mg/L of D-catechin) were analyzed following the method described by Swain and Hillis [31]. Orthodiphenols (OD, as mg/L of D-catechin) were analyzed by Paronetto [29]. The ionization index (ION-I) was analyzed by Somers and Evans method [32] and gelatin index (GEL-I), ethanol index (EtOH-I) and hydrochloric acid index (HCl-I) by Ribéreau-Gayón [33].

2.6.2. Color Analysis

Color intensity was determined by measuring absorbance at 420, 520, and 620 nm in a 1 mm cell. Other variables calculated were red, yellow, and blue percentages, according to Glories [34].

Spectral readings (transmittance every 10 nm over the visible spectrum, 380–770 nm, and absorbance measurements at 420, 520, and 620 nm) were performed with a PerkinElmer's LAMBDA 25 UV/vis Spectrophotometer (Waltham, MA, USA), using 1 mm path length cuvette. All the parameters were measured in duplicate in every sample.

2.6.3. Copigmentation Parameter Determination

Copigmentation was determined according to the method proposed by Boulton [35], via the following parameters, where the color was due to (TA) total anthocyanins; (COP) copigmentation; (AL) free anthocyanins; (PP) polymeric pigment; (FC) the estimation of the content of flavanol cofactors; and (TP) the estimation of the content of total phenols (monomers and tannins).

2.6.4. Anthocyanin Individual Determination

Anthocyanins were analyzed by HPLC-DAD according to del Alamo Sanza et al. [8], as mg/L of malvidin-3-*O*-glucoside: delphinidin-3-*O*-glucoside (Df-3-Gl), cyanidin-3-*O*-glucoside (Cy-3-Gl), petunidin-3-*O*-glucoside (Pt-3-Gl), peonidin-3-*O*-glucoside (Pn-3-Gl), malvidin-3-*O*-glucoside (Mv-3-Gl), vitisin A (vitA); acetyl derivates: peonidin-3-*O*-acetylglucoside (Pn-3-Gl-Ac) and malvidin-3-*O*-acetylglucoside (Mv-3-Gl-Ac); coumaryl derivates: delphinidin-3-*O*-*p*-coumarylglucoside (Df-3-Gl-Cm), cyanidin-3-*O*-*p*-coumarylglucoside (Cy-3-Gl-Cm), petunidin-3-*O*-*p*-coumarylglucoside (Pt-3-Gl-Cm), malvidin-3-*O*-*p*-coumarylglucoside *cis*-C and *trans*-T (Mv-3-Gl-Cm); and ethyl-linked malvidin-3-*O*-glucoside-ethyl-epicatechin (Mv-3-gl-Ethyl).

2.7. Statistical Analysis

Correlation coefficients and principal component analysis were performed using the Statgraphics Centurion XVII statistical program (version XVII; StatPoint, Inc., Warrenton, VA, USA).

3. Results and Discussion

3.1. Micro-Oxygenation Strategy and Evolution of Oxygen Consumption in Wines

The two micro-oxygenation (MOX) strategies selected would be approximated: the normal published oak barrel oxygen ingress rates (50 µg/L) [13,26] and another higher one (3 mL/L·month). Throughout the aging period (111 days), continuous measurements of dissolved oxygen (µg/L) were carried out in each tank. Figure 2 shows the evolution of this dissolved oxygen in wine for each combination of MOX/wood strategy, together with the dosage applied in each one. The difference between a floating and fixed wine MOX can be observed immediately: in the first (Figure 2A1,A2), the dosage varies according to the measure of dissolved oxygen, the wine's initial level being 0.5 mL/L·month. Each day after exhaustive measuring, as described in the previous section, the level was adjusted to reach the fixed setpoint (50 µg/L). In the second set of cases (Figure 2B1,B2) the measurements were also carried out, but the oxygen dosage was the same throughout the 111 days of the experiment: 3 mL/L·month.

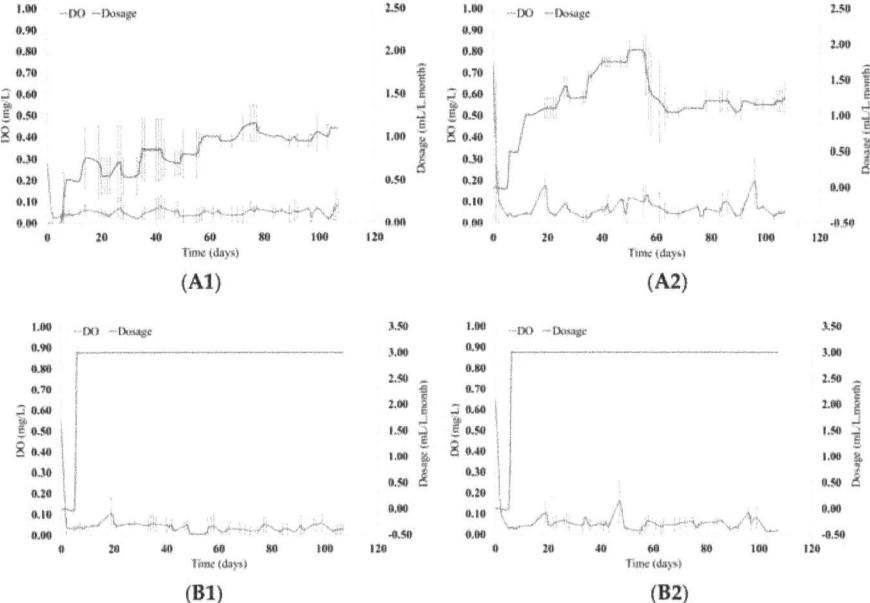

Figure 2. Graphic representation of dissolved oxygen (mg/L) and MOX dosage (mL/L·month) of wines treated with two micro-oxygenation strategies: (**A**) floating MOX strategy, in which the setpoint was set at 50 μg/L; (**B**) fixed MOX strategy with 3 mL/L·month. The wood chips were added at two different moments: (**1**) all the wood at the beginning of the maturation process; (**2**) half of the wood at the beginning and the other half 48 days after the beginning of the experiment.

The average trends in the evolution of accumulated oxygen consumption in wines treated with oak chips from *Q. Pyrenaica*, added at two different times during aging with different MOX strategies, is shown in Figure 3. The data obtained indicate that the wines treated during 111 days with a floating MOX strategy, in which the setpoint was fixed at 50 μg/L, and all the oak chips were added at the beginning (from now on, A wood strategy) consumed an average of 5.86 ± 0.473 mg/L with significant differences from the wines to which the wood was added twice (from now on, B wood strategy), which consumed an average of 7.87 ± 0.287 mg/L. The wines treated with a fixed MOX strategy of 3 mL/L·month and A wood strategy were recorded as consuming an average of 16.58 ± 0.017 mg/L without any significant differences from the wines to which the oak chips were added according to the B wood strategy, and which consumed an average of 16.59 ± 0.001 mg/L.

From the start point (Figure 3, detail number 1), the aging systems were noticeably different in accordance with both the micro-oxygenation and wood dosage strategies. In wines treated with fixed MOX, the most obvious difference appeared from the start point until the moment the second quantity of oak chips was added (Figure 3, detail number 2). No significant differences were observed among them after that time. The average dosage during the whole process was the same for both: 2.04 ± 0.00 mg/L·month (Table 1). With regard to the floating MOX strategy, differences were appreciable between tanks from the start: B wood strategy wines needed a higher oxygen dosage to maintain the setpoint levels than those to which all the wood was added at the beginning of the process (Figure 3, detail number 3). Therefore, the average dosages during the whole process were 0.83 ± 0.24 mg/L·month and 0.62 ± 0.14 mg/L·month (Table 1), respectively. Related to the total oxygen inputs, no significant differences were recorded among the values of wines with a fixed MOX (16.60 ± 0.02 and 16.41 ± 0.25 mL/L, Table 1). However, in the case of those treated with a floating

MOX, wines aged with the B wood strategy received over 1.50 mL/L more oxygen than wines treated with the A wood strategy (Table 1).

Table 1. Summary of dosages during MOX in the aging tanks tested.

Micro-Oxygenation	50 µg/L		3 mL/L·month	
Time wood was added	At the beginning	Half of the wood at the beginning and the other half at 48 days	At the beginning	Half of the wood at the beginning and the other half at 48 days
MOX dosage (mL/L·month)	0.92 ± 0.17	1.22 ± 0.35	3.00 ± 0.00	3.00 ± 0.00
Total O$_2$ inputs (mL/L)	6.06 ± 0.27	7.60 ± 0.37	16.60 ± 0.02	16.41 ± 0.25
MOX dosage (mg/month)	0.62 ± 0.14	0.83 ± 0.24	2.04 ± 0.00	2.04 ± 0.00

Total O$_2$ inputs include MOX + O$_2$ from the air inside the wood chips and the oxygen contributed during initial tank filling.

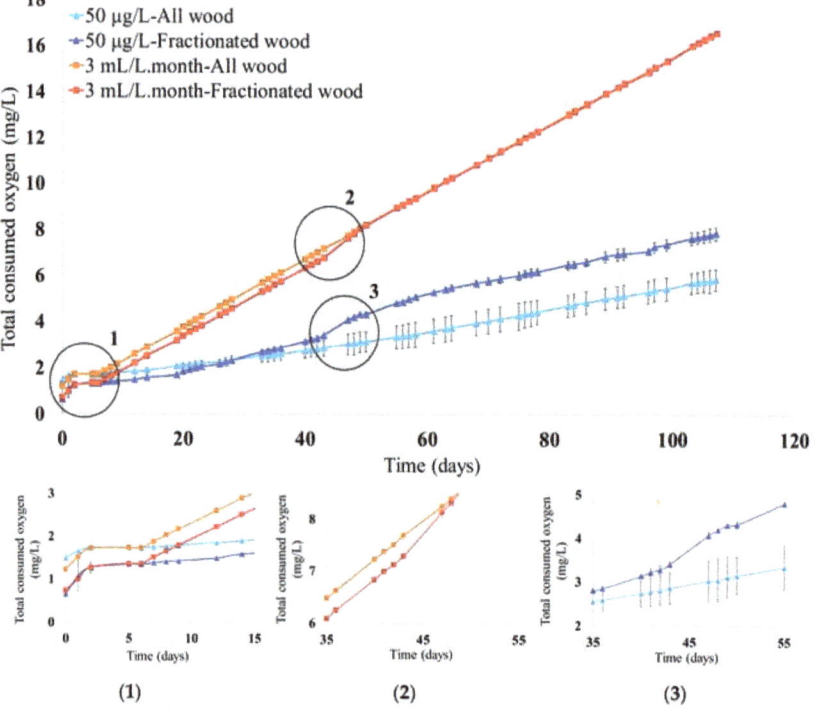

Figure 3. Graphic representation of MOX total consumption of oxygen (TCO mean, mg/L) in wines treated with two micro-oxygenation strategies. ▲: floating MOX strategy (50 µg/L) and all chips added at the beginning of the experiment; ▲: floating MOX strategy (50 µg/L) and half of the chips added at the beginning and the other half 48 days later; ■: fixed MOX strategy (3 mL/L·month) and all chips added at the beginning of the experiment; ■: fixed MOX strategy (3 mL/L·month) and half of the chips added at the beginning and the other half 48 days later.

These results showed that wines consumed all the oxygen available. Also, if a MOX strategy was followed for the wines to retain the amounts of oxygen similar to that in barrel (20 to 50 µg/L) [13,26], between 5 and 8 mg/L of oxygen was required to maintain those levels. However, when they were micro-oxygenated with the fixed dosages common in finished wines, for example 3 mL/L·month, this was between 3 and 3.5 times more than the oxygen they would receive in barrel. When wood was added at the beginning, the micro-oxygenated wines, with a fixed dosage of 3 mL/L·month, consumed 2.8 times more oxygen than those subjected to floating MOX. Similarly, when the wood was added twice, the micro-oxygenated wines with a fixed dosage of 3 mL/L·month consumed 2.10 times more than those subjected to floating MOX.

The contribution of the oxygen from the air entrapped in the oak chips (auto-oxygenation) was evident when both wood and MOX strategies were taken into account. On the one hand, the oxygen contributed by the wood was appreciable: in those tanks where the A wood strategy was tested, it was higher than those where the B wood strategy was used. On the other hand, the oxygen contributed by the oak chips had a greater repercussion when the floating MOX strategy was used (Figure 3, detail number 3): 13.08% and 9.75% of the total oxygen inputs for wood strategies A and B, respectively. In wines treated with fixed MOX (Figure 3, detail number 2) this contribution was 4.62%. Hence, under the conditions tested in this study, the oxygen contribution from the oak chips was higher when a lower MOX was used, in this case, an FMOX. It should be noted that when wine consumes the required quantity of oxygen, auto-oxygenation needs to be taken into account, since it is an important part of the total oxygen consumed, as previously stated. When an FMOX is considered, logically, it has less influence, since it overlaps with the fixed dosage. However, in both cases, the oxygen contained in the wood should be considered as it acts as an oxidizing agent of the compounds released by the wood [14].

3.2. Effect of Micro-Oxygenation in Wines

The measurements carried out throughout the aging process showed that the factors involved (MOX and time of oak chip addition) did not generally affect the chemical wine parameters at the end of aging, with the exception of the degree of alcohol, since wines from floating MOX and A wood strategy had a significantly higher value with respect to the rest of the wines (data not shown). The final chemical parameters of these wines varied in range: total acidity 5.85 to 5.90 g/L; volatile acidity 0.45–0.49 g/L; sugars 1.35–1.39 g/L; degree of alcohol 13.43–14.05%; and the pH of all the wines studied were close to 3.69.

The oxygen consumed by the wine determines its evolution during the period of contact with wood + MOX, thus explaining the final wine differences. Correlations between total consumed oxygen (TCO) by wine and each variable analyzed in the wines were calculated and are shown in Table 2. They were calculated with TCO and the variation of each parameter in each sample, calculated with respect to the value at the initial time (Delta). The parameters studied were (a) parameters of phenols (D-LPP, D-HPP, D-TAN, D-CAT, D-ACY, D-OD); (b) copigmentation indexes (D-COP, D-PP, D-AL, D-FC, D-TP D-HCl-I, D-EtOH-I, D-Ion-I, D-GEL-I); (c) color parameters (D-T, D-%A420, D-%A520, D-%A620); and d) individual anthocyanin compounds (D-Df-3-Gl, D-Cy-3-Gl, D-Pt-3-Gl, D-Pn-3-Gl, D-Mv-3-Gl, D-Vitisin A, D-Pn-3-Gl-Ac, D-Mv-3-Gl-Ac, D-Df-3-Gl-Cm, D-Cy-3-Gl-Cm, D-Pt-3-Gl-Cm, D-Mv-3-Gl-Cm C, D-Mv-3-Gl-Cm T and D-Mv-3-gl-Ethyl). The correlation values significant with total consumed oxygen at $p < 0.05$ ($r > 0.51$ or $r < -0.51$) are in bold (Table 2), where the positive correlation indicates an increase in compound concentration and the negative one a decrease. Significant correlations ($p < 0.05$) were recorded between TCO and some of the parameters studied in wines treated with both MOX strategies. Table 2 showed the expected increase of D-HPP, which correlated significantly with TCO in wines from both MOX strategies, and had similar values (0.4928 and 0.4708, for 50 µg/L and 3 mL/L·month, respectively). Anthocyanin (D-ACY) evolution correlated negatively to oxygen consumption reflect the anthocyanin decrease. A great number of their reactions during wine aging are known to be determined by DO level [36,37]. This negative correlation was higher in the case of wines with a fixed MOX (−0.6219), which can be explained by higher quantities of anthocyanins lost in wines submitted to this MOX strategy. The anthocyanin decrease is one of the most important processes occurring during wine storage and was previously reported by Del

Alamo et al. [13] when a 20 µg/L FMOX was carried out with different alternative oak products. These processes encompass oxidation, condensation and phenolic polymerization with transformation into other compounds which have an evident effect on the modification of wine color. The correlation results shown by the red component (D-%A520) agreed with the above: TCO showed a significant negative correlation with this color component in wines, with a higher value in those from floating MOX. Also, the previous results were in agreement with the free anthocyanins (D-AL) value from copigmentation parameters, since TCO showed a statistically significant ($p < 0.05$) correlation with its decrease, but only in wines from a floating MOX (−0.6023). In the same way, the D-ION-I was negatively correlated with TCO, especially in the case of wines treated with fixed MOX (−0.7754), which coincided with the higher negative value in the D-ACY parameter for these wines.

Table 2. Mean and standard deviation of each delta analyzed parameter, and correlation coefficients between parameters and oxygen consumed by wine in each micro-oxygenation and wood strategies process.

Parameter	50 µg/L			3 mL/L·month		
	Means	Std. Dev.	TCO (µg/L)	Means	Std. Dev.	TCO (µg/L)
TCO (µg/L)	4731.553	2027.881	1.000000	10834.07	4902.537	1.000000
D-LPP	15.865	92.833	0.141855	27.19	173.186	0.439508
D-HPP	575.850	182.728	**0.492817**	497.68	247.238	**0.470771**
D-CAT	5.581	116.711	−0.138876	−12.84	70.835	−0.380093
D-ACY	−106.595	62.470	**−0.521031**	−90.65	33.655	**−0.621881**
D-TAN	3.026	0.285	0.118416	2.92	0.187	0.078329
D-CI	−1.789	3.240	**−0.561610**	−1.99	3.315	**−0.683212**
D-T	0.115	0.075	**0.782966**	0.12	0.077	**0.909374**
D-FC	−0.400	1.666	−0.098521	−0.61	1.681	−0.261112
D-TP	−11.231	14.193	−0.453367	−10.55	12.323	−0.321815
D-COP	0.029	0.048	0.065126	0.04	0.056	0.097685
D-AL	−0.154	0.032	**−0.602250**	−0.16	0.065	−0.286415
D-PP	0.074	0.040	0.402556	0.07	0.034	0.381631
D-EtOH-I	6.613	2.660	−0.431844	5.36	4.294	**−0.658603**
D-HCl-I	3.529	2.668	**−0.464282**	3.50	3.059	−0.488752
D-GEL-I	−20.827	7.877	−0.095990	−19.63	5.599	−0.029458
D-ION-I	−8.443	8.122	**−0.516031**	−9.93	8.022	**−0.775361**
D-OD	−182.100	60.460	**−0.738996**	−162.80	53.456	**−0.756095**
D-IPT	0.103	6.066	0.009464	−1.37	3.020	0.002513
D-%A420	0.975	3.415	0.318594	0.92	3.513	0.379195
D-%A520	−7.347	5.000	**−0.588424**	−7.58	5.243	**−0.645097**
D-%A620	6.372	6.934	0.267373	6.66	7.274	0.281831
D-Df-3-Gl	−4.957	6.689	−0.359792	−5.45	6.259	−0.164040
D-Cy-3-Gl	−2.007	1.846	**−0.876709**	−2.00	1.808	**−0.829679**
D-Pt-3-Gl	−2.196	3.909	−0.162831	−2.37	3.643	0.055472
D-Pe-3-Gl	−0.883	1.095	−0.407683	−0.92	1.005	−0.226321
D-Mv-3-Gl	−10.421	11.659	−0.436685	−10.85	10.893	−0.264131
D-Vitisin A	0.179	0.191	**0.657446**	0.18	0.193	**0.659883**
D-Mv-3-Gl-Ethyl	0.108	0.205	**0.540355**	0.09	0.213	**0.699438**
D-Pe-3-Gl-Ac	−0.485	0.523	−0.350600	−0.59	0.440	−0.323565
D-Df-3-Gl-Cm	−2.470	1.986	**−0.813771**	−2.52	1.918	**−0.868531**
D-Mv-3-Gl-Ac	−0.258	0.190	−0.361746	−0.30	0.203	−0.107868
D-Cy-3-Gl-Cm	0.954	1.845	0.114454	1.02	1.962	0.130637
D-Mv-3-Gl-Cm C	0.316	0.336	**0.525493**	0.42	0.575	0.383817
D-Pt-3-Gl-Cm	−0.152	0.214	−0.467845	−0.14	0.164	−0.390148
D-Mv-3-Gl-Cm T	−2.396	1.985	**−0.875855**	−2.36	1.944	**−0.884408**
D-Acet	4.270	6.419	**0.677614**	3.63	5.936	**0.771172**
D-Cum	5.378	8.351	**0.660566**	5.44	8.005	**0.810550**
D-Total	32.185	76.224	**0.516223**	30.13	73.811	**0.678868**

Significant correlation values ($r > 0.51$ or $r < −0.51$). Bold type indicates at least $p < 0.05$.

The tonality (D-T) increase in wines correlated positively with TCO, especially in the case of those treated with fixed MOX (0.9094). The negative correlations present in color intensity (D-CI) were higher in wines from a fixed MOX strategy (-0.6832) than those from a floating strategy (-0.5616). Del Alamo et al. [13] reported a positive correlation between TCO and D-CI when a 20 µg/L FMOX was studied.

TCO showed significant correlations with some individual anthocyanins. While D-Cy-3-Gl, D-Df-3-Gl-Cm and D-Mv-3-Gl-Cm T content defined wines in the first sampling, their decrease throughout the aging process correlated negatively with TCO, with values similar in wines from each MOX studied (Table 2). In wines from floating MOX, D-Pt-3-Gl-Cm also correlated negatively with TCO. Contrary to the D-Mv-3-Gl-Cm T, the *cis* isomer correlated positively with TCO (0.5255) in wines from floating MOX. D-vitisin A correlated positively ($p < 0.05$) with the TCO increase, with values very close for both MOX strategies: 0.6574 and 0.6599 for floating and fixed MOX strategies, respectively. Vitisin A, as a pyranoanthocyanin, is an important compound in the color of red wines, since the cycloaddition process strongly increases product stability. In this way, vitisin A has been reported as being more stable than Mv-3-Gl or ethyl-linked compounds, and more resistant to oxidation [38].

Related to ethyl-linked compounds, D-Mv-3-Gl-Ethyl increased significantly in wines throughout the aging period. This compound, purple in color, is less sensitive to bleaching by SO_2 and pH than monomeric anthocyanins, and its formation is favored by oxygen [36], as shown by the positive correlation between this compound and TCO.

Figures 4–6 present the results of principal component analysis (PCA) of the variation in variables analyzed in wines (Delta). This analysis was carried out to obtain a reduced number of linear combinations of the variables that explain the greater variability in the data. The projections of the variables analyzed in the principal components (PCs) are the weighted sum of the original variables and are named loads (Table 3). Using the variables of total oxygen consumed, phenolic compounds and color parameters, 3 components with eigenvalues greater than or equal to 1.0 were obtained. They explain 74% of the variability in the original data, where the first main component included 34.6%, the second 27.6%, and the third 11.73% (Table 3 PCA-A). Projection of the variables on the factor-plane (1×2) (Figure 4A) and of the wines on the factor-plane, show that wines were located according to their aging time (Figure 4B), and demonstrate the significance of the variables in the samples with aging time. There was a greater distance between the samples of the first and second sampling, which indicates a greater evolution between 20 and 48 days, while there was almost no differentiation between 76 and 97 days of aging. Finally, the separation between the samples after 111 days of aging was remarkable, indicating different characteristics of the wines aged in the different systems according to the variables studied. The first main component PC1 contains, on the one hand, information on oxygen consumption and tonality and, on the other, the red component related to the loss of anthocyanins. The second main component PC2 is primarily defined by the yellow component information, which relates to the formation of polymerized phenolic compounds on the one hand, and the blue color component on the other. According to the distribution of the samples the youngest wines, after one month's aging, were located in the negative PC1 and logically defined by the free anthocyanins, catechins, and orthodiphenols, showing significant levels of compounds responsible for their red color. As aging progresses, after 3 months' contact with wood and MOX, the wines had significant levels of compounds responsible for blue tones, which in the following month, changed to brown, showing the importance of the yellow color component. Wines at the end of the period of contact with wood (after 111 days) were located in the positive PC1 and defined by their consumption of oxygen (TCO), which is directly related to the formation of highly polymerized phenolic compounds, and increase in wine tonality and the yellow component (Figure 4B).

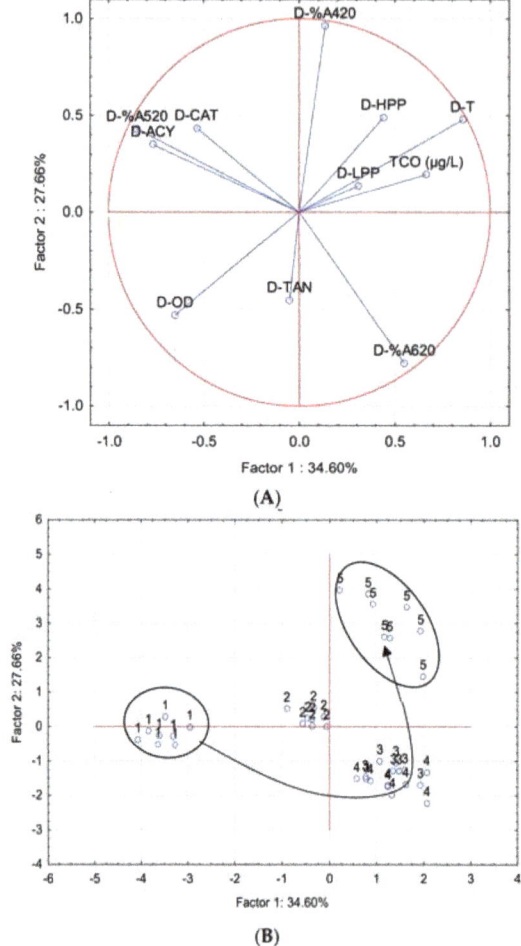

Figure 4. Principal component analysis (PCA) performed with global phenol parameters in wines from different micro-oxygenation and wood strategies. 1, 2, 3, 4, and 5: samplings carried out in each tank after 20, 48, 76, 97, and 111 days' aging, respectively.

As regards the capacity of wine differentiation by accumulated oxygen consumption (TCO), copigmentation, phenolic indexes, and color Delta parameters (Figure 5), the three main components comprised 77.40% of the variance (Table 3 PCA-B), where the first and second main components explained 31.6% and 25.41%, respectively, whereas the third was 20.39%. As before, wines were shown based on their aging time according to the projection of the variables on the factor-plane (1 × 2) (Figure 5A) and the projection of the wines on the factor-plane (Figure 5B). In this case, the wines of the first two samplings were very close, which indicates low evolution of the variables studied between 20 and 48 days. Similar trends were observed in samplings 3 and 4. The first main component, PC1, encompasses positive variables, such as ionization index and the blue color component, whereas the yellow component and gelatin index were negative. Meanwhile, the second main component PC2 was mainly defined by the information on oxygen consumption relating to the loss of anthocyanins and represented by two parameters: the red color component and free anthocyanins from the copigmentation parameters. The distribution of the samples showed that after one month's aging, wines were located in the negative

PC2, and defined by the anthocyanins, both as free ones and as a percentage which contributed to color. They also showed significant levels of compounds responsible for the red color, as previously stated (Figure 4A). As the wine aging progressed, after 3 months' contact with wood and MOX, wines were found to show significant levels of compounds responsible for blue tones. In the following months, these changed to brown, showing the importance of the yellow color component (after 111 days). At the end of the aging period, wines were located in the positive PC2, and were defined by their consumption of oxygen (TCO), which was directly related to the copigmentation parameter (COP) and the astringent tannin content through the gelatin index (Figure 5B).

Table 3. PCA results.

Parameter	Factor 1	Factor 2
PCA-A		
TCO (µg/L)	0.6634	0.1964
D-LPP	0.3082	0.1378
D-HPP	0.4402	0.4895
D-CAT	−0.5360	0.4341
D-ACY	−0.7675	0.3499
D-TAN	−0.0502	−0.4529
D-T	0.8553	0.4782
D-%A420	0.1334	0.9629
D-%A520	−0.8512	0.4295
D-%A620	0.5486	−0.7790
D-OD	−0.6510	−0.5311
PCA-B		
TCO (µg/L)	−0.1588	0.7599
D-%A420	−0.9154	0.1802
D-%A520	−0.5577	−0.7365
D-%A620	0.8483	0.4431
D-COP	−0.4493	0.2756
D-TP	0.5143	−0.4087
D-FC	0.5851	−0.2164
D-AL	0.0251	−0.6377
D-PP	0.5921	0.4860
D-EtOH-I	0.1834	−0.6141
D-HCl-I	0.0113	−0.5556
D-GEL-I	−0.7125	−0.2527
D-ION-I	0.7163	−0.5213
PCA-C		
TCO (µg/L)	0.3890	0.6159
D-%A420	−0.3931	0.5658
D-%A520	−0.8618	−0.2132
D-%A620	0.8129	−0.1221
D-Df-3-Gl	−0.9192	0.2025
D-Cy-3-Gl	−0.5871	−0.7686
D-Pt-3-Gl	−0.8454	0.4356
D-Pe-3-Gl	−0.9341	0.1583
D-Mv-3-Gl	−0.9427	0.1100
D-Vitisin A	−0.0175	0.9354
D- Mv-3-Gl-Ethyl	−0.1916	0.9326
D-Pe-3-Gl-Ac	−0.8297	0.1714
D-Df-3-Gl-Cm	−0.7212	−0.5033
D-Mv-3-Gl-Ac	−0.3799	−0.1920
D-Cy-3-Gl-Cm	0.6353	−0.4160
D-Mv-3-Gl-Cm C	0.5347	0.0076
D-Pt-3-Gl-Cm	−0.8031	−0.0258
D-Mv-3-Gl-Cm T	−0.4527	−0.8697

Figure 6 summarizes the principal component analysis carried out with individual anthocyanin compounds and color parameters, which revealed that the three main components explained 82.97% of the variance, where the first main component includes 46.13% and the second 25.63% (Table 3 PCA-C). Figure 6A,B include the projection of the variables and the wine samples in the plane of the first two main components respectively, showing that wines were located according to their aging time, as was also found with the previously studied parameters (Figures 4 and 5). On the one hand, information related to the yellow component was included in the first main component PC1 and, on the other hand, the red component related to the loss of individual anthocyanins, where compounds such as malvidin-3-O-glucoside, peonidin-3-O-glucoside, and delphinidin-3-O-glucoside are worth noting, in this order. The second main component, PC2, was mainly defined by vitisin A and malvidin-3-O-glucoside-ethyl-epicatechin, along with information on oxygen consumption and the loss of malvidin-3-O-p-coumarylglucoside *trans*, and cyanidin-3-O-glucoside. Considering the distribution of wines throughout the aging process, they were located in the negative PC2 from the first month of aging to the third one. They were defined by compounds responsible for the red color and anthocyanin monomers (malvidin-3-O-glucoside, peonidin-3-O-glucoside, and delphinidin-3-O-glucoside) in the younger wines. Sample 2 showed variations in malvidin-3-O-p-coumarylglucoside *trans*, and cyanidin-3-O-glucoside until the blue color component and anthocyanins, such as cyanidin-3-O-p-coumarylglucoside and malvidin-3-O-p-coumarylglucoside *cis*, were recorded in wines from the second month. Wines at the end of the period of contact with wood and MOX, which covers the fourth and fifth sampling, were located in the positive PC2 and defined by the consumption of oxygen (TCO) and anthocyanin derivates more stable to oxidation, such as vitisin A and malvidin-3-O-glucoside-ethyl-epicatechin (Figure 6B), showing significant levels of the compounds responsible for yellow tones.

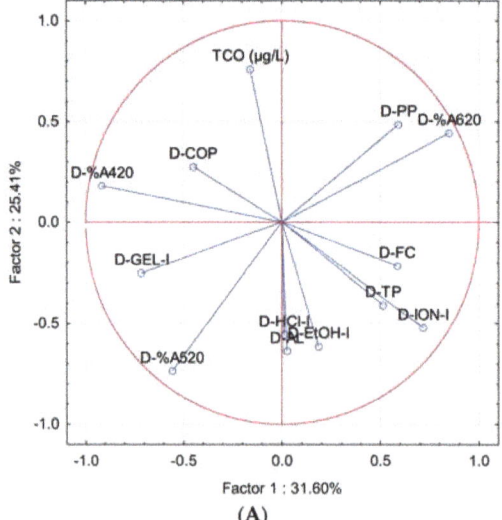

Figure 5. *Cont.*

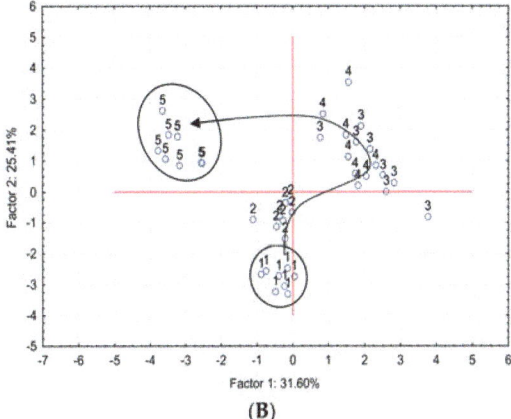

Figure 5. Principal component analysis (PCA) performed with copigmentation parameters in wines from different micro-oxygenation and wood strategies. 1, 2, 3, 4, and 5: samplings carried out in each tank after 20, 48, 76, 97, and 111 days' aging, respectively.

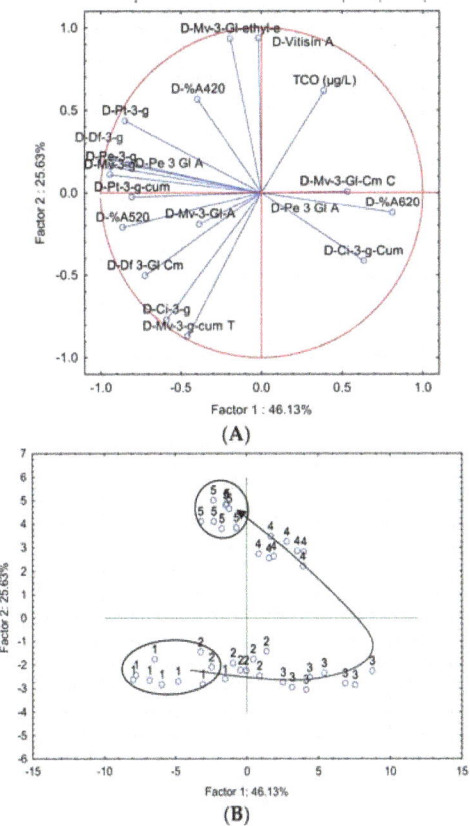

Figure 6. Principal component analysis (PCA) performed with individual anthocyanin compounds in wines from different micro-oxygenation and wood strategies. 1, 2, 3, 4, and 5: samplings carried out in each tank after 20, 48, 76, 97, and 111 days' aging, respectively.

4. Conclusions

The results of oxygen consumption by red wines under different micro-oxygenation (MOX) strategies and *Q. pyrenaica* chips showed that they consumed all the oxygen available, and that this consumption depended on the MOX strategy selected. Thus, wines from fixed MOX received between 3 and 3.5 times more oxygen than those using the floating MOX strategy. In the latter, the oxygen contribution by the air entrapped in the wood was more significant, since this contribution represented a higher percentage with respect to the total. Therefore, the oxygen contribution from the chips and the moment when this contribution is made should be taken into account, because it affects the MOX strategy selected. In general, although the amount of oxygen that had been applied and the way it was applied is different, it does not appear that these are important differences in the studied time. However, it will be interesting to continue the study of the wines in the bottle in order to evaluate the effect of this great difference in the oxygen consumed by the wines. According to this information we could say that the wines are similar, even though we believe that it is more appropriate to implement floating dosage MOX, that is, to give the wine the oxygen it needs at every moment.

In relation to their effect on color and phenolic characteristics, both MOX strategies studied contributed to the characteristic processes which take place during barrel aging: oxygen consumption was related to copigmentation increment, with the consequent loss of monomeric anthocyanins together with an increase in yellow tones, related to the formation of polymerized phenolic compounds. From the point of view of individual global phenol and copigmentation parameters, 111 days' aging was the period where a higher evolution was observed, while in case of individual anthocyanin compounds, there was hardly any evolution between the last two sampling points.

Consequently, the use of chips combined with MOX is an adequate technique to carry out the aging process in tanks.

Author Contributions: Conceptualization, I.N. and M.d.A.-S.; Funding acquisition, I.N. and M.d.A.-S.; Methodology, M.d.A.-S.; Supervision, I.N.; Visualization, R.S.-G.; Writing—original draft, R.S.-G. and A.M.M.-G.; Writing—review & editing, R.S.-G., I.N. and M.d.A.-S.

Funding: This research was funded by the Ministry of Economy and Competitiveness-FEDER of the Spanish Government for Project AGL2014-54602-P, Junta de Castilla y León for project VA028U16 and Interreg Spain-Portugal for Iberphenol project.

Acknowledgments: R.S.-G. would like to thank her postdoctoral contract to Interreg Spain-Portugal for Iberphenol project. The authors wish to thank to J. Calles for her support throughout the chemical analysis. Authors thank Ann Holliday for her services in revising the English.

Conflicts of Interest: The authors declare no conflict of interest.

References

1. Anli, R.E.; C avuldak, Ö.A. A review of microoxygenation application in wine. *J. Inst. Brew.* **2012**, *118*, 368–385. [CrossRef]
2. Chiciuc, I.; Farines, V.; Mietton-Peuchot, M.; Devatine, A. Effect of wine properties and operating mode upon mass transfer in micro-oxygenation. *Int. J. Food Eng.* **2010**, *6*, 9. [CrossRef]
3. Du Toit, W.J.; Marais, J.; Pretorius, I.S.; du Toit, M. Oxygen in must and wine: A review. *S. Afr. J. Enol. Vitic.* **2006**, *27*, 76–94. [CrossRef]
4. Parish, M.; Wollan, D.; Paul, R. Micro-oxygenation—A review. *Aust. N.Z. Grapegrow. Winemak.* **2000**, *438a*, 47–50.
5. Parpinello, G.P.; Versari, A. Micro-oxygenation of red wine: Chemistry and sensory aspects. In *Wine: Types, Production and Health*; Peeters, A.S., Ed.; Nova Science Publishers: New York, NY, USA, 2012; pp. 93–123. ISBN 978-161470635-9.
6. Trivellin, N.; Barbisan, D.; Badocco, D.; Pastore, P.; Meneghesso, G.; Meneghini, M.; Zanoni, E.; Belgioioso, G.; Cenedese, A. Study and development of a fluorescence based sensor system for monitoring oxygen in wine production: The WOW project. *Sensors* **2018**, *18*, 1130. [CrossRef] [PubMed]

7. Del Alamo Sanza, M.; Nevares Domínguez, I.; Cárcel Cárcel, L.M.; Navas Gracia, L. Analysis for low molecular weight phenolic compounds in a red wine aged in oak chips. *Anal. Chim. Acta* **2004**, *513*, 229–237. [CrossRef]

8. Del Álamo Sanza, M.; Domínguez, I.N.; Merino, S.G. Influence of different aging systems and oak woods on aged wine colour and anthocyanin composition. *Eur. Food Res. Technol.* **2004**, *219*, 124–132. [CrossRef]

9. Del Alamo-Sanza, M.; Laurie, V.F.; Nevares, I. Wine evolution and spatial distribution of oxygen during storage in high-density polyethylene tanks. *J. Sci. Food Agric.* **2014**, *95*, 1313–1320. [CrossRef] [PubMed]

10. Oberholster, A.; Elmendorf, B.L.; Lerno, L.A.; King, E.S.; Heymann, H.; Brenneman, C.E.; Boulton, R.B. Barrel maturation, oak alternatives and micro-oxygenation: Influence on red wine aging and quality. *Food Chem.* **2015**, *173*, 1250–1258. [CrossRef] [PubMed]

11. Gallego, L.; Del Alamo, M.; Nevares, I.; Fernández, J.A.; De Simón, B.F.; Cadahía, E. Phenolic compounds and sensorial characterization of wines aged with alternative to barrel products made of Spanish oak wood (*Quercus pyrenaica* Willd.). *Food Sci. Technol. Int.* **2012**, *18*, 151–165. [CrossRef] [PubMed]

12. Nevares, I.; Del Alamo, M.; Cárcel, L.M.; Crespo, R.; Martin, C.; Gallego, L. Measure the dissolved oxygen consumed by red wines in aging tanks. *Food Bioproc. Technol.* **2009**, *2*, 328–336. [CrossRef]

13. Del Álamo, M.; Nevares, I.; Gallego, L.; Fernández de Simón, B.; Cadahía, E. Micro-oxygenation strategy depends on origin and size of oak chips or staves during accelerated red wine aging. *Anal. Chim. Acta* **2010**, *660*, 92–101. [CrossRef] [PubMed]

14. García-Estévez, I.; Alcalde-Eon, C.; Martínez-Gil, A.M.; Rivas-Gonzalo, J.C.; Escribano-Bailón, M.T.; Nevares, I.; del Alamo-Sanza, M. An approach to the study of the interactions between ellagitannins and oxygen during oak wood aging. *J. Agric. Food Chem.* **2017**, *65*, 6369–6378. [CrossRef] [PubMed]

15. Nevares, I.; del Álamo-Sanza, M. Wine aging technologies. In *Recent Advances in Wine Stabilization and Conservation Technologies*; Jordão, A.M., Cosme, F., Eds.; Nova Science Publishers: New York, NY, USA, 2016; pp. 209–245, ISBN 978-163484899-2; 978-163484883-1.

16. De Simón, B.F.; Cadahía, E.; Sanz, M.; Poveda, P.; Perez-Magariño, S.; Ortega-Heras, M.; González-Huerta, C. Volatile compounds and sensorial characterization of wines from four Spanish denominations of origin, aged in Spanish Rebollo (*Quercus pyrenaica* Willd.) oak wood barrels. *J. Agric. Food Chem.* **2008**, *56*, 9046–9055. [CrossRef] [PubMed]

17. Castro-Vázquez, L.; Alañón, M.E.; Ricardo-Da-Silva, J.M.; Pérez-Coello, M.S.; Laureano, O. Study of phenolic potential of seasoned and toasted Portuguese wood species (*Quercus pyrenaica* and *Castanea sativa*). *J. Int. Sci. Vigne Vin* **2013**, *47*, 311–319. [CrossRef]

18. De Coninck, G.; Jordão, A.M.; Ricardo-Da-Silva, J.M.; Laureano, O. Evolution of phenolic composition and sensory properties in red wine aged in contact with Portuguese and French oak wood chips. *J. Int. Sci. Vigne Vin* **2006**, *40*, 25–34. [CrossRef]

19. De Simón, B.F.; Sanz, M.; Cadahía, E.; Poveda, P.; Broto, M. Chemical characterization of oak heartwood from Spanish forests of *Quercus pyrenaica* (Wild.). Ellagitannins, low molecular weight phenolic, and volatile compounds. *J. Agric. Food Chem.* **2006**, *54*, 8314–8321. [CrossRef] [PubMed]

20. De Simon, B.F.; Cadahia, E.; Jalocha, J. Volatile compounds in a Spanish red wine aged in barrels made of Spanish, French, and American oak wood. *J. Agric. Food Chem.* **2003**, *51*, 7671–7678. [CrossRef] [PubMed]

21. De Simón, B.F.; Muiño, I.; Cadahía, E. Characterization of volatile constituents in commercial oak wood chips. *J. Agric. Food Chem.* **2010**, *58*, 9587–9596. [CrossRef] [PubMed]

22. Del Alamo Sanza, M.; Escudero, J.A.F.; De Castro Torío, R. Changes in phenolic compounds and colour parameters of red wine aged with oak chips and in oak barrels. *Food Sci. Technol. Int.* **2004**, *10*, 233–241. [CrossRef]

23. Castellari, M.; Simonato, B.; Tornielli, G.B.; Spinelli, P.; Ferrarini, R. Effects of different enological treatments on dissolved oxygen in wines. *Ital. J. Food Sci.* **2004**, *16*, 387–396.

24. Moutounet, M.; Mazauric, J.-P. L'oxygène dissous dans les vins. *Rev. Fran. D'Oenol.* **2001**, *186*, 12–15. Available online: https://prodinra.inra.fr/record/56312 (accessed on 20 June 2018).

25. Laurie, V.F.; Law, R.; Joslin, W.S.; Waterhouse, A.L. In situ measurements of dissolved oxygen during low-level oxygenation in red wines. *Am. J. Enol. Vitic.* **2008**, *59*, 215–219.

26. Nevares, I.; del Álamo, M. Measurement of dissolved oxygen during red wines tank aging with chips and micro-oxygenation. *Anal. Chim. Acta* **2008**, *621*, 68–78. [CrossRef] [PubMed]

27. Folin, O.; Ciocalteu, V. On tyrosine and tryptophane determination in proteins. *J. Biol. Chem.* **1927**, *73*, 627–650.

28. Masquelier, J.; Michaud, J.; Triaud, J. Fractionnement des leucoanthocyannes du vin. *Bull. Soc. Pharm. Bordx.* **1965**, *104*, 81–85.

29. Paronetto, L. *Polifenoli e Tecnica Enológica*; Edagricole: Bologna, Italy, 1977; ISBN 88-206-1704-8.

30. Ribereau-Gayon, J.; Stonstreet, E. Determination of anthocyanins in red wine. In *Bulletin de la Societe de Pharmacie de Bordeaux*; Société de Pharmacie (Bordeaux): Bordeaux, France, 1965; Volume 9, pp. 2649–2652. ISSN 00378968.

31. Swain, T.; Hillis, W.E. The phenolic constituents of Prunus domestica. I.—The quantitative analysis of phenolic constituents. *J. Sci. Food Agric.* **1959**, *10*, 63–68. [CrossRef]

32. Somers, T.C.; Evans, M.E. Spectral evaluation of young red wines: Anthocyanin equilibria, total phenolics, free and molecular SO_2, "chemical age". *J. Sci. Food Agric.* **1977**, *28*, 279–287. [CrossRef]

33. Ribéreau-Gayón, P.; Glories, Y.; Maujean, A.; Dobourdieu, D. *Tratado de Enología 2. Química del vino. Estabilización y Tratamientos*; Ediciones Mundi prensa: Madrid, España, 2003; ISBN 978950504571.

34. Glories, Y. La couleur des vins rouges 2. Mesure, origine et interprétation. *Connais. Vigne Vin* **1984**, *18*, 253–271. [CrossRef]

35. Boulton, R.B. The copigmentation of anthocyanins and its role in the colour of red wine: A critical review. *Am. J. Enol. Vitic.* **2001**, *52*, 67–87.

36. Atanasova, V.; Fulcrand, H.; Cheynier, V.; Moutounet, M. Effect of oxygenation on polyphenol changes occurring in the course of wine-making. *Anal. Chim. Acta* **2002**, *458*, 15–27. [CrossRef]

37. Castellari, M.; Matricardi, L.; Arfelli, G.; Galassi, S.; Amati, A. Level of single bioactive phenolics in red wine as a function of the oxygen supplied during storage. *Food Chem.* **2000**, *69*, 61–67. [CrossRef]

38. Morata, A.; Gomez-Cordoves, C.; Colomo, B.; Suarez, J.A. Pyruvic acid and acetaldehyde by different strains of Saccharomyces cerevisiae: Relationship with vitisin A and B formation in red wines. *J. Agric. Food Chem.* **2003**, *51*, 7402–7409. [CrossRef] [PubMed]

Article

Use of Oak Fragments during the Aging of Red Wines. Effect on the Phenolic, Aromatic, and Sensory Composition of Wines as a Function of the Contact Time with the Wood

Pilar Rubio-Bretón, Teresa Garde-Cerdán and Juana Martínez *

Instituto de Ciencias de la Vid y del Vino (Gobierno de La Rioja, CSIC, Universidad de La Rioja),
Carretera de Burgos Km. 6, 26007 Logroño, Spain; rubio_pilar@hotmail.com (P.R.-B.);
teresa.garde@icvv.es (T.G.-C.)
* Correspondence: jmartinezg@larioja.org; Tel.: +34-941-894-980

Received: 31 October 2018; Accepted: 29 November 2018; Published: 5 December 2018

Abstract: The use of oak fragments allows wine cellars to reduce costs and the length of wine aging compared to traditional aging in oak barrels in the winery. The main objective of this work was to study the effect of the use of oak fragments on the volatile, phenolic, and organoleptic characteristics of Tempranillo red wines, as a function of the contact time between the wood and the wine. The results showed important changes in the wines' colorimetric parameters after two months of contact time. Extraction kinetics of volatile compounds from the wood was highest during the first month of contact for chips, variable for staves, and slower and continuous over time for barrels. Wines macerated with fragments showed the best quality in short periods of aging, while barrel-aged wines improved over the time they spent in the barrel. In addition, the results allowed an analytical discrimination between the wines aged with oak fragments and those aged in oak barrels, and between chips and staves, just as at the sensory level with triangular tasting tests. In conclusion, the use of oak fragments is a suitable practice for the production of red wines, which may be an appropriate option for wines destined to be aged for short periods.

Keywords: oak fragments; oak barrels; volatile compounds; phenolic compounds; sensory analysis; triangular tasting

1. Introduction

The aging of red wines in oak barrels is a technique commonly used in wineries to increase wine stability and complexity. During this process, an organoleptic improvement of the wines is achieved as a consequence of the contribution of oak wood compounds, and the phenolic and aromatic modifications that take place [1,2]. This practice involves long aging periods and represents a high economic cost for the wineries.

The use of pieces of oak wood during winemaking, as an alternative to traditional aging in oak barrels, is an enological practice authorized by the International Vine and Wine Organisation (OIV) and included in the International Enological Codex [3,4]. This practice was approved by the European Community [5], and is subject to regulation [6,7].

Currently, there is a varied range of commercial products available, and therefore their effects on wine quality can be very variable, since they are influenced by numerous factors (size of fragments, oak wood origin, toasting degree, manufacturing process, dose, contact time with wine, etc.) [8,9]. Due to the large contact surface of these materials with the wine, the extraction of compounds is much faster than in barrels. Additionally, the cost of the process is lower than that of the classic aging in barrels.

The shape and size of the oak fragments available vary: powder, chips, pieces of medium size (cubes or beans, dominoes, blocks or segments), or larger pieces (staves) to put in the tanks. These products are made with oak from different oak origins (American, French, Spanish, Hungarian, etc.), using different toasting processes (direct fire, convection by hot air, or infrared radiation), and different levels of toasting (medium, strong, light, or untoasted). These effects have been largely investigated, but the influence of the length of the aging process needs further studies.

Depending on the characteristics of the final product desired, the contact time of the wood with the wine can vary from a few days to several weeks, and even months. This contact time will depend on the type of wood, the fragment size, the dose, and the sensory profile expected to be achieved in the wines.

Most previous studies about the influence of the length of the aging process have dealt with a particular subgroup of compounds (volatile or phenolic), which is clearly a limiting condition to obtain a more comprehensive view of the subject. The main objective of this work was to study for the first time the evolution of the aromatic and phenolic composition of wine during the contact time with oak wood, employing four different accelerated aging strategies: chips and staves, with and without micro-oxygenation. In addition, the impact of these treatments on the wine organoleptic characteristics was evaluated. These aging procedures were compared with the traditional oak barrels aging method.

2. Materials and Methods

2.1. Wine and Wood

To carry out this study, a red wine of the Tempranillo variety (*Vitis vinifera* L.) was used. Once the malolactic fermentation (MLF) was finished, the wine was divided into eight stainless steel tanks of 250 L capacity and two 225 L oak barrels. In four tanks, oak chips were added at a dose of 4 g/L, while staves were placed in the other four. In this case, the dose was adjusted so as to have a contact surface with the wine of 0.4 m^2/hL. All the fragments were of American oak with medium toasting. Simultaneously, two of the tanks with chips and two with staves were micro-oxygenated with a dose of 2 mL/L/month during the first two months, and 1 mL/L/month during the following two months. Thus, the total dose of oxygen during 4 months was 6 mL/L. Micro-oxygenation treatments were carried out using a Micro-Ox-3V system (Intec, Verona, Italia) connected to the stainless steel tanks. These tanks (Herpanor S.A., Laguardia, Spain) were 2 m in height and a diameter of 0.45 m. The dimensions of the tanks were chosen because of the necessity to achieve the height recommended for micro-oxygenation, and to reproduce the height to diameter ratio of common red wine tanks used in industrial-scale production. These dimensions were necessary so that the small oxygen bubbles produced during micro-oxygenation would have a sufficient displacement height to guarantee their complete dissolution into the wine. Oxygen was provided through a diffuser composed of a porous ceramic membrane placed 10 cm above the bottom of the tank. The contact period of the oak fragments with the wine was 6 months, and then, the wines were bottled to be stored for 18 months. In addition, the same wine was simultaneously aged for 12 months in American oak barrels with medium toasting, manufactured in the same cooperage as the chips and staves. After this aging period, the wine was bottled to be bottle-aged for another year (Figure 1). In this way, all the tests were carried out in duplicate.

The chemical parameters of the wine before aging were: alcoholic strength 13.7% *v*/*v*, pH 3.81, total acidity 4.88 g/L, and volatile acidity 0.51 g/L. These parameters were evaluated using the analytical methods established by the European Community [10].

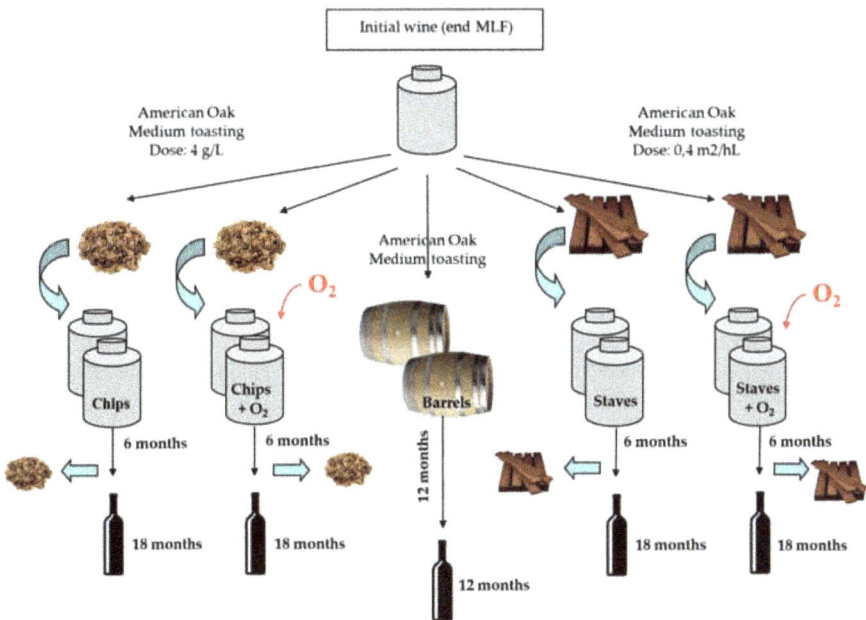

Figure 1. Experimental design.

2.2. Chemical Analysis

2.2.1. Color Parameters Measurements

Color intensity (CI) was determined according to the EEC Regulation 2676/90 [10]. Percentage of yellow, red, and blue components (% yellow, % red, and % blue, respectively) were evaluated according to the methodology described by Glories [11]; while the total polyphenol index (TPI) and total anthocyanins were determined by the methods described by Ribéreau-Gayon et al. and Ribéreau-Gayon and Stonestreet, respectively [1,12]. Color parameter measurements were made before starting the treatments, and after 2, 4, 6, 12, and 24 months.

2.2.2. Wine Volatile Compounds from Oak Wood

The analysis of the volatile compounds in wine coming from oak wood was carried out by Gas Chromatography (GC), with a method based on that described by Ortega et al. [13], under optimized conditions. To carry out the extraction of the volatile compounds, in a glass tube were added: 5 mL of wine, previously centrifuged at 4000 rpm at 0 °C for 10 min; 9.5 mL of supersaturated ammonium sulfate solution; 15 µL of a solution of 2-octanol and 3,4-dimethylphenol (internal standards) in ethanol at a concentration of 50 mg/L; and 0.2 mL of dichloromethane. These tubes were shaken vigorously, first manually, and then horizontally in an orbital agitator at 400 rpm for 60 min. Next, the tubes were centrifuged at 2500 rpm and 0 °C for 10 min. The supernatant aqueous phase was removed with a Pasteur pipette and the organic phase was collected. This extract was transferred to a microtube and centrifuged at 13,000 rpm and 0 °C for 5 min, to break up any possible emulsions formed. Finally, the organic phase was collected with a syringe and transferred to a vial with an insert, in order to analyze in the gas chromatograph.

The separation and detection of the compounds was carried out in a gas chromatograph HP-6890 series II equipped with an automatic injector and a flame ionization detector (FID). The column was a DB-WAX (50 m × 0.20 mm × 0.2 µm thick film, J and W Scientific, Folsom, CA, USA), using nitrogen as a carrier gas with a flow rate of 0.6 mL/min. The injection was carried out in splitless mode (0.5 min) with an injection volume of 2 μL. The chromatographic conditions used were the following: injector temperature, 250 °C; detector temperature, 275 °C (H_2 flow, 40 mL/min; air flow, 450 mL/min; auxiliary gas, N_2 at 40 mL/min); and initial oven temperature, 75 °C (5 min), 3.7 °C/min to 240 °C (maintained for 15 min), and a post time of 10 min at 240 °C.

The identification of the volatile compounds was performed by comparison with the retention times of the standard substances, and the concentration of each substance was measured by comparing it with calibrations made with the pure compounds analyzed under the same conditions.

Wine volatile compounds from oak wood analysis were made before starting the treatments, and after 1, 2, 4, 6, 12, and 24 months.

2.2.3. Low Molecular Weight Phenols

Low molecular weight phenols were analyzed by HPLC-DAD with direct injection of the 30 μL of sample, according to Martínez and Rubio-Bretón [14]. A column Zorbax Eclipse Plus C18 (300 mm × 150 mm × 3.9 μm) was used. The eluents used in the mobile phase were: A (water/acetic acid, 98/2, *v/v*), B (water/acetonitrile/acetic acid, 78/20/2, *v/v/v*), and C (methanol), with a constant flow of 0.9 mL/min according to the following program: 0 to 80% of B from 0 to 65 min; 80% B from 65 to 85 min; 100% C from 86 to 90 min. The wine samples were centrifuged (4000 rpm/0 °C/10 min) and filtered by 0.45 μm before injection into the equipment. The identification of the compounds was carried out by comparing the retention times and the spectral parameters of the chromatographic peaks with those of the standards.

The wine was analyzed before starting the treatments, and after 1, 2, 4, 6, 12, and 24 months.

2.3. Sensorial Analysis

The organoleptic study was carried out during the process at 2, 4, 6, 12, and 24 months. Ten wine tasters carried out the sensory analysis, all of them with extensive experience as wine tasters. The samples were evaluated comparatively by a blind tasting system and served in a random order. A sufficient amount of wine samples were presented without any identification according to regulation UNE 87-022-92 [15]. All the wines were served at room temperature and were evaluated in individual booths. The score sheet used was based on that used in some wine competitions and is considered official in some designations of origin. According to this model sheet the wines are evaluated based on the absence of defects, so the lower the score, the higher the quality of the wine. The sensory attributes valued were visual, olfactory and taste phases, and harmony. Quantitative evaluation of aromatic descriptors was also carried out on an intensity scale of 1 to 10 (fruity, varietal, spices, wood-toasted, almond-caramel, vanilla, and smoked) as well as the taste characteristics (structure, persistence, retronasal aroma, and astringency).

Triangular tasting tests were also carried out, at 6 and 12 months, to determine the existence of detectable differences between two different samples, according to regulation UNE 87-006-92 [15]. To conduct these tests, three coded samples were presented simultaneously to the tasters, two of which were the same, so that the taster could identify which was the different sample. Although the aim of this technique is not the determination of preferences, the preferred sample in each series was also specified. The percentage of preferences was calculated with the correct answers, discarding the preferences of the incorrect tests.

2.4. Statistical Analysis

Canonical discriminant analysis (CDA) was performed with the concentrations of volatile and phenolic compounds in the different samples for all the moments studied. The "IBM SPSS Statistics 22" statistical program was used.

3. Results and Discussion

3.1. Color Parameters of Wine

Figure 2 shows the evolution of the main colorimetric parameters over aging time. In general, all the samples followed a similar chromatic evolution, independent of the treatment carried out.

CI of the wines increased during the first two months, mainly in the micro-oxygenated wines and in the wines aged in barrels. From that moment, this parameter decreased in all the treatments, maintaining the initial differences until the end of the aging process (Figure 2a). These results coincide with those obtained by other authors [16,17]. Furthermore, a higher CI in micro-oxygenated wines was also observed by other authors [18,19], which could be due to an increase in the blue color by the contribution of pigments with ethyl bridges. As noted by these authors, a pronounced increase in the blue color percentage in micro-oxygenated wines was observed (Figure 2f).

In the same way as CI, the percentage of red color increased in all wines until two months, with a notable decrease observed from that moment (Figure 2d). Del Álamo et al. [20] also observed a loss of CI, especially in the red component, from the third month of aging with fragments and barrels.

The percentage of yellow color in wines followed a trend similar to the tonality (data not shown), appreciating a decrease during the first two months, and then a significant increase until the end of the period studied (Figure 2b). Pérez-Prieto et al. [21] attributed the increase in yellow tones to the extraction of color compounds from the oak wood during aging. Our results also coincided with those of Cadahía et al. [22], who observed a decrease in the percentage of red color and an increase in the percentage of yellow color and tonality, while these authors did not observe significant variations in the CI in wines aged for 12 months in oak barrels.

A decrease in TPI was observed over the aging period (Figure 2c). This parameter, at 4 and 12 months of aging, was higher in barrels compared to the other aging systems, but at 24 months, the highest TPI value corresponded to wine with staves and micro-oxygenation. In the case of total anthocyanins, the decrease was more pronounced in micro-oxygenated wines during the first 12 months, although these differences disappeared at 24 months (Figure 2e). Tavares et al. [23] also observed a decrease in anthocyanins during wine contact with chips, probably due to anthocyanin condensation and polymerization reactions, and the precipitation of these compounds during wine aging.

3.2. Volatile Compounds from Oak Wood in Wine

Figure 3 shows the evolution of furanic compounds, benzoic aldehydes, and oak lactones during the period of aging. Furanic compounds, with the exception of furfuryl alcohol, are formed during wood toasting through degradation of carbohydrates [24]. Wines in contact with staves and aged in barrels had a much higher concentration of these compounds than those treated with chips (Figure 3a–c). Moreover, their evolution was similar, with their content increasing during the first 6 months and then decreasing sharply, until practically disappearing in the case of furanic aldehydes.

Towey and Waterhouse [25] also observed that the concentration of these compounds decreased significantly after 7 months of barrel aging. Garde-Cerdán and Ancín-Azpilicueta [26] also found a decrease of furfural and 5-methylfurfural from the sixth month of aging in new barrels, as well as 5-hydroxymethylfurfural from the ninth month. Other authors [27] also observed a maximum extraction in furfural at 6 months of barrel aging and a decrease from that moment. The degradation of these compounds can be due to their reduction to the corresponding alcohols by biological mechanisms [28], although they can also participate in other reactions that contribute to the decrease of their concentrations in free form in wines (such as the formation of 2-furanmethanethiol [29] or of color adducts with the wine catechin [30]). For a short time of aging, extraction of these compounds from the wood is greater than their degradation so that they are accumulated in wine. However, when the aging time increases, degradation reactions may exceed the extraction, so that their concentration tends to decrease [24,31,32].

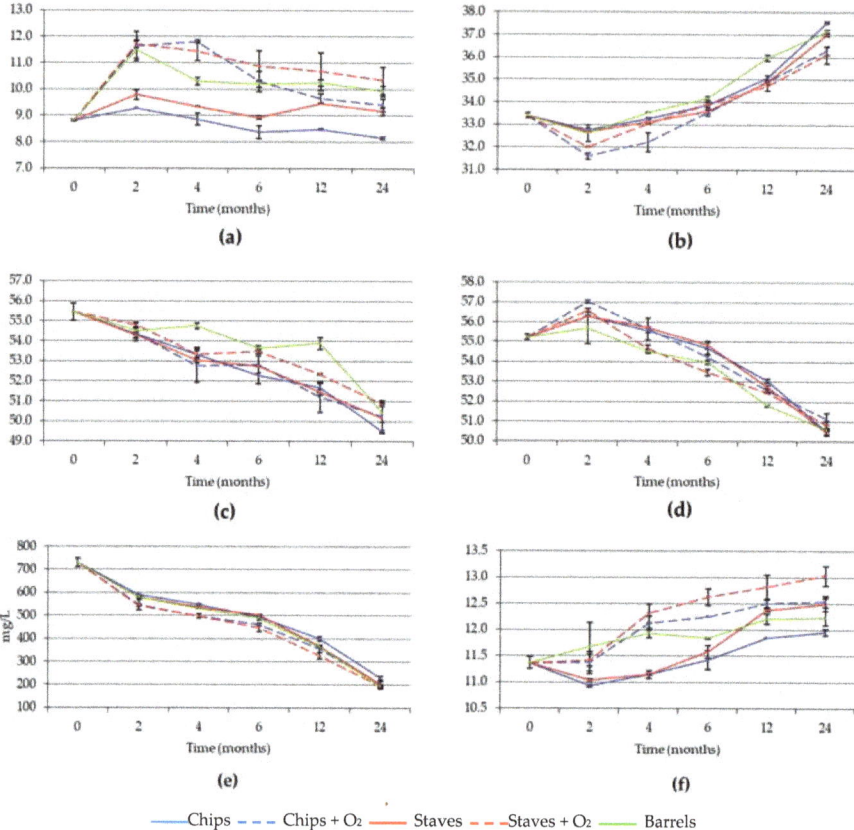

Figure 2. Evolution of the main chromatic parameters over time: (**a**) color intensity (CI); (**b**) percentage of yellow color; (**c**) total polyphenol index (TPI); (**d**) percentage of red color; (**e**) total anthocyanins (mg/L); and (**f**) percentage of blue color. Mean ± standard deviation ($n = 2$).

On the other hand, in wines treated with oak chips, hardly any extraction of furfural or 5-methylfurfural was observed, while the content of 5-hydroxymethylfurfural increased during the first month of contact with the wood and remained practically constant until 6 months, at which point it disappeared. Similar results were obtained by Fernández de Simón et al. [33], who observed the highest amount of furanic aldehydes in wines treated with chips at 30 days, with the concentration of these also being lower than those treated with staves.

The concentration of furfuryl alcohol also reached higher values in wines treated with staves and aged in barrels (Figure 3d). This compound originates from the microbiological reduction of furfural, even after the alcoholic and malolactic fermentations have been completed [34], and its concentration depends on factors that affect enzymatic reactions, such as pH, temperature, or residual microbiological activity [35].

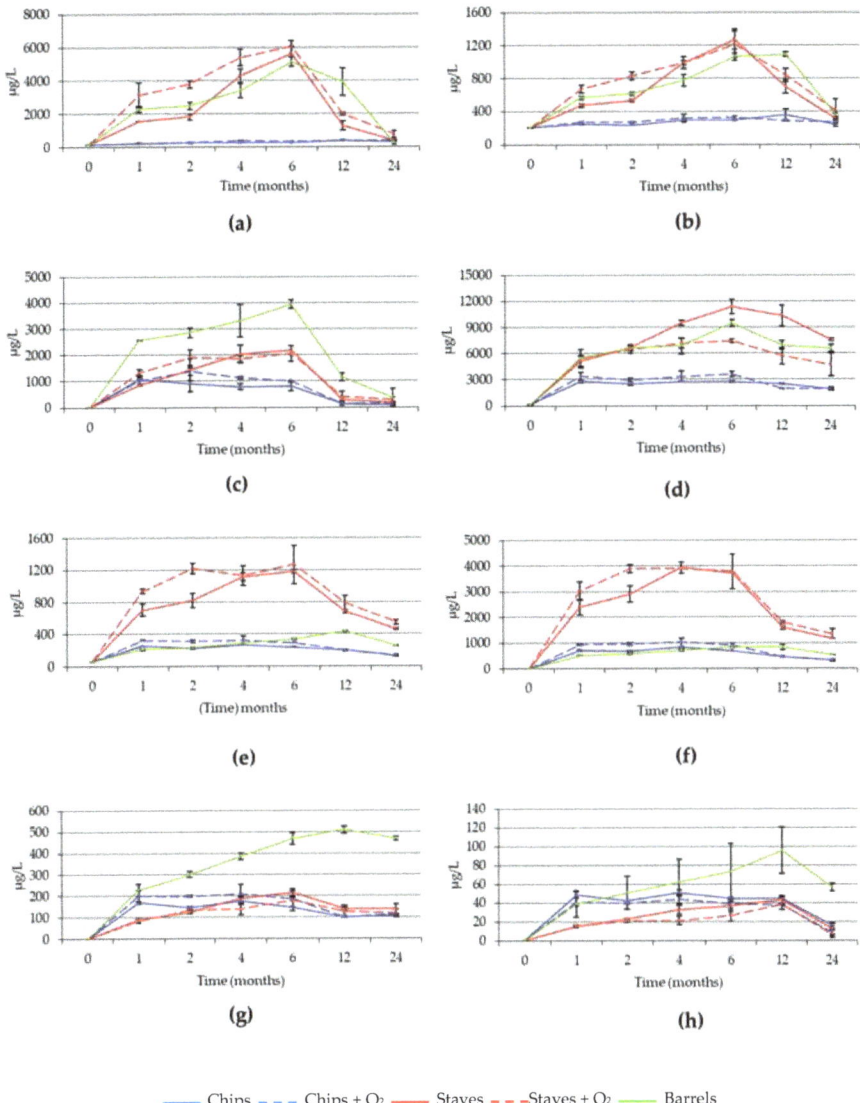

Figure 3. Evolution over time of furanic compound concentration (µg/L): (**a**) furfural, (**b**) 5-methylfurfural, (**c**) 5-hydroxymethylfurfural, and (**d**) furfuryl alcohol; of benzoic aldehyde concentration (µg/L): (**e**) vanillin, and (**f**) syringaldehyde; and of oak lactone concentration (µg/L): (**g**) *cis*-β-methyl-γ-octalactone, and (**h**) *trans*-β-methyl-γ-octalactone. Mean ± standard deviation (*n* = 2).

Benzoic aldehydes (vanillin and syringaldehyde) are formed during the thermal degradation of lignin. These compounds achieved their maximum concentration during the first month in wines with oak chips, and between 2 and 4 months in the case of wines treated with staves, which obtained the highest concentrations throughout the process studied (Figure 3e–f). In both cases, the concentration decreased from 6 months, probably due to the microbiological reduction to the corresponding alcohols [31,36,37]. On the other hand, the wines aged in barrels presented a lower content of these compounds than the wines in contact with staves, increasing until 12 months of aging, and decreasing later during the bottle-aging period. Coinciding with our results, different authors [24,32,38] found that the concentration of benzoic aldehydes was at maximum after 10–12 months of barrel aging. Like furanic aldehydes, vanillin accumulates in wine during short aging times, since initially the extraction is high due to the different concentration between wine and wood [26]. However, when the aging time is prolonged, it can be transformed into vanillic alcohol, so that the vanillin concentration can decrease.

In wines treated with oak chips, the maximum extraction of the two isomers of β-methyl-γ-octalactone occurred during the first month of contact with the wood. In the case of staves, *cis*-β-methyl-γ-octalactone was extracted during the 6 months of contact time, while the *trans* isomer was extracted up to 12 months (Figure 3g–h). Other authors [39–41] also observed that the accumulation of *cis* and *trans* oak lactones increased with the aging time of wines in barrels. Meanwhile, wines aged in barrels showed a higher content of these compounds than those macerated with fragments, with these two isomers increasing during the 12 months of contact with the wood, and then decreasing during bottle-aging. The decrease in the concentration of these compounds could be due to the wine undergoing different chemical transformations [42].

The evolution of volatile phenols during the aging time is shown in Figure 4. These compounds are formed from the thermal degradation of the lignin at a high temperature. In general, a greater extraction of these compounds was observed in the wines in contact with the staves, mainly in the case of 4-methylguaiacol, *trans*-isoeugenol, and syringol.

Guaiacol concentration increased throughout the process, obtaining higher values in wines treated with staves and in those aged in barrels, reaching its maximum concentration at 12 months of aging, at which point its content remained practically constant (Figure 4a). These results coincide with those obtained by Garde-Cerdán et al. [38]. The extraction kinetics of 4-methylguaiacol showed major differences between treatments, finding the highest concentration in wines in contact with staves at 6 months, while in wines aged with chips and in oak barrels, it was achieved during the first month of contact (Figure 4b). Pérez-Prieto et al. [43] observed that 4-methylguaiacol reached its maximum concentration at 3 months, while the guaiacol extraction continued for up to a maximum of 9 months of aging in barrels. Although the staves had medium toasting, their appearance indicated that they had undergone a stronger toasting than the chips, probably because the toasting system applied was different. This fact could explain the higher content of guaiacol and 4-methylguaiacol in wines in contact with staves, since they are compounds that are formed at high temperatures of toasting [26]

As can be seen in Figure 4c,d, the wines aged in barrels sharply increased the levels of ethylphenols (4-ethylguaiacol and 4-ethylphenol) during their aging in bottles. These compounds can be extracted from wood in very low concentrations, but mainly they are formed during the aging of wines by microbiological transformations of cinnamic acids carried out by *Brettanomyces/Dekkera* yeast contaminants [44]. The concentrations of ethylphenols found in wines aged in barrels at the end of the process exceeded those considered harmful to the wine aroma, 140 and 620 µg/L for 4-etilguaiacol and 4-ethylphenol, respectively [45]. This fact could be due to contamination of the wines aged in barrels during the bottling process.

The concentration of 4-vinylguaiacol tended to decrease progressively up to 12 months, when the concentration in wines with chips and staves remained practically constant (Figure 4e). However, in wines aged in barrels, its concentration decreased sharply, probably as a consequence of its reduction to 4-ethylguaiacol by the effect of contamination by *Brettanomyces/Dekkera*. The 4-Vinylphenol content decreased slightly or remained constant in all the wines up to 12 months. In the case of wines aged in barrels, its concentration remained practically constant from this moment, and yet, in wines in contact with chips and staves, its content increased until 24 months (Figure 4f).

The concentration of phenol increased until 12 months in all treatments (Figure 4g). From that moment, its concentration continued to increase in wines in contact with staves, while its level decreased slightly in wines aged with chips and more sharply in wines aged in oak barrels. Garde-Cerdán et al. [38]

found the highest concentration of this compound at 10 months of aging in barrels and they also observed a decrease in its content from that moment.

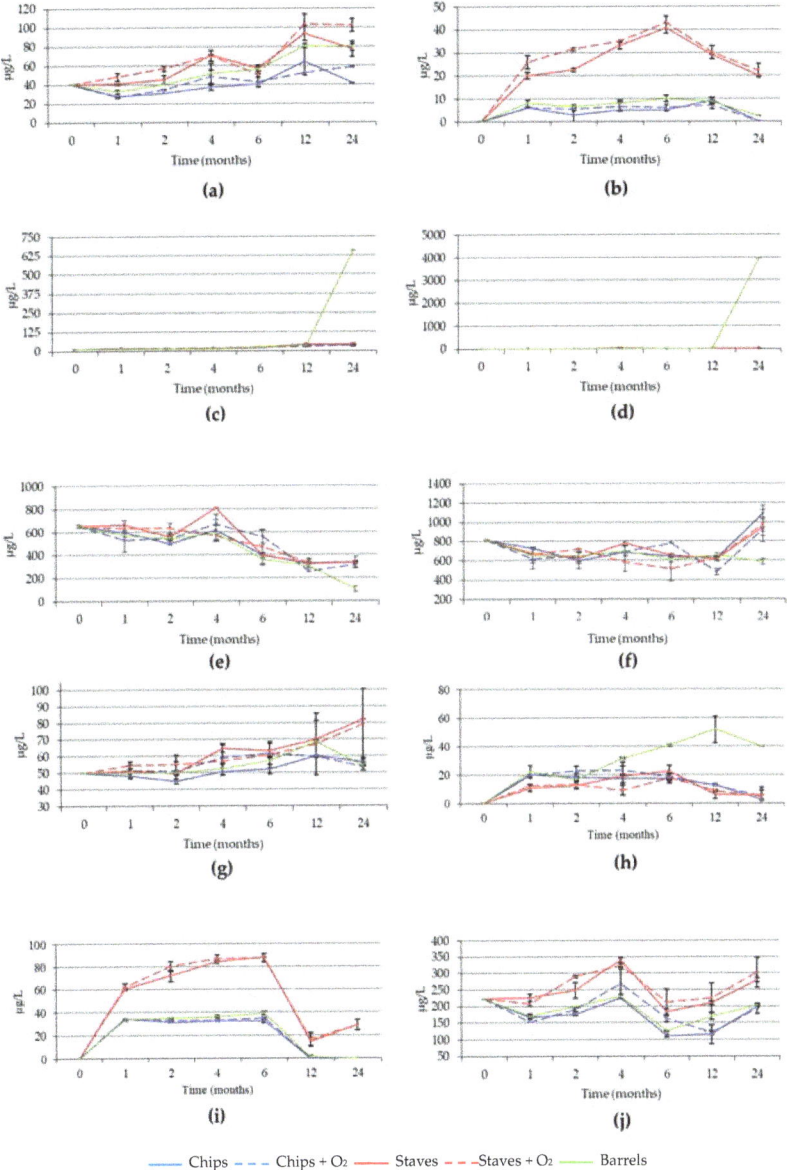

Figure 4. Evolution of volatile phenols concentration (µg/L) over time: (**a**) guaiacol; (**b**) 4-methylguaiacol; (**c**) 4-ethylguaiacol; (**d**) 4-ethylphenol; (**e**) 4-vinylguaiacol; (**f**) 4-vinylphenol; (**g**) phenol; (**h**) eugenol; (**i**) *trans*-isoeugenol; and (**j**) syringol. Mean ± standard deviation (*n* = 2).

On the other hand, the maximum extraction of eugenol was reached in wines during the first month of contact with chips, and after 6 months in the wines macerated with staves. In addition, eugenol was extracted constantly during the contact time of wines with the oak barrels, thus obtaining a maximum concentration at 12 months, and its concentration was much higher than that obtained in wines in contact with fragments (Figure 4h). Similar results regarding the increase in barrels of this compound were observed by other authors [25,37,40,41].

The *trans*-Isoeugenol reached its maximum concentration in the first month of contact in wines aged in the barrel and with chips, and at 4–6 months in those treated with staves, which was much higher than in the other wines. In all the wines, a decrease in the concentration of this compound was observed after 6 months, disappearing in wines aged in oak barrels and with chips (Figure 4i). Finally, syringol presented a similar evolution in all the treatments, reaching the maximum concentration at 4 months, which was higher in wines treated with staves throughout the process (Figure 4j). Different results were obtained by Fernández de Simón et al. [33], who observed an increase in the concentrations of *trans*-isoeugenol and syringol until the end of aging in wines treated with alternative oak products.

3.3. Low Molecular Weight Phenols

Benzoic acids are constituents of wood, and its content in wine increased as a result of contact with it (Figure 5). These results are in agreement with those observed by other authors [46,47].

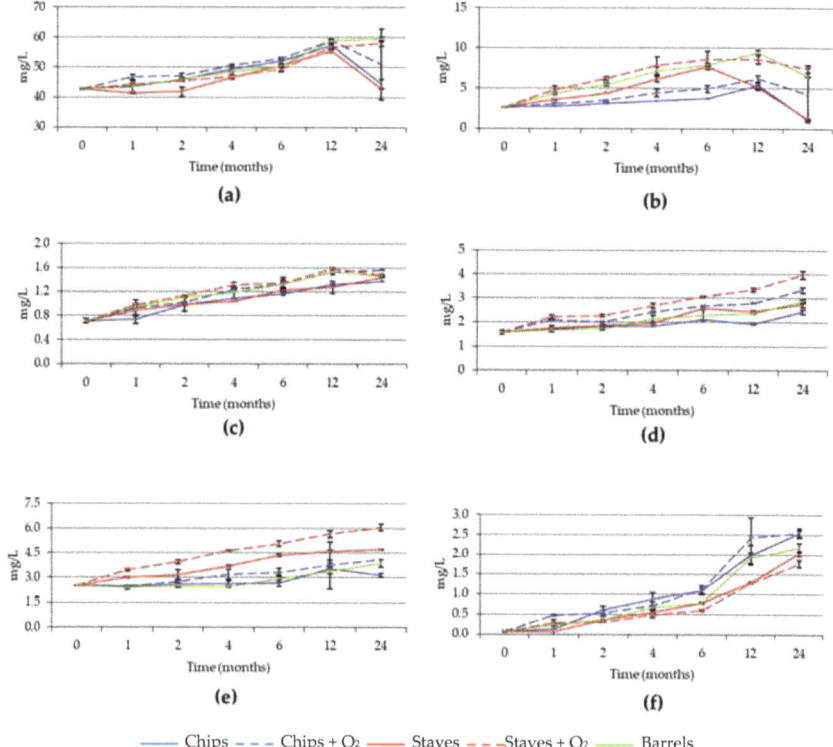

Figure 5. Evolution of benzoic acids concentration (mg/L) over time: (**a**) gallic acid; (**b**) protocatechuic acid; (**c**) p-hydroxybenzoic acid; (**d**) vanillic acid; (**e**) syringic acid; and (**f**) ellagic acid. Mean ± standard deviation (*n* = 2).

Gallic and ellagic acids are very important compounds due to their strong antioxidant activity, even at very low concentrations [48]. The concentration found for both compounds was slightly higher in wines treated with chips in most of the process (Figure 5a,f), probably due to its greater contact surface and a less intense toasting than in the staves, since gallic acid is a compound which is sensitive to thermal degradation. Similar results were obtained by Alañón et al. [46] when comparing chestnut wood chips and barrels.

Protocatechuic acid concentration increased more in wines in contact with staves and in wines aged in barrels than in those treated with chips at almost every moment (Figure 5b). This may be because this compound is generated during the toasting process, as a consequence of the thermal degradation of lignin [49], and as has been noted previously the toasting seemed more intense in the staves than in the chips.

Higher concentrations of p-hydroxybenzoic and vanillic acids were detected in micro-oxygenated wines (Figure 5c,d), as well as syringic acid in wines treated with staves, compared to those aged in the barrel and in contact with chips (Figure 5e). The content of these three compounds increased in the wines during the entire study period.

Cinnamic acids were detected in the wines in their free forms (caffeic, coumaric, and ferulic acids) and in their respective tartaric esters (caftaric, coutaric, and fertaric acids). The concentration of these acids in wine depends on grape variety and winemaking technique, but they are not found in oak wood. The evolution of their content over time is shown in Figure 6.

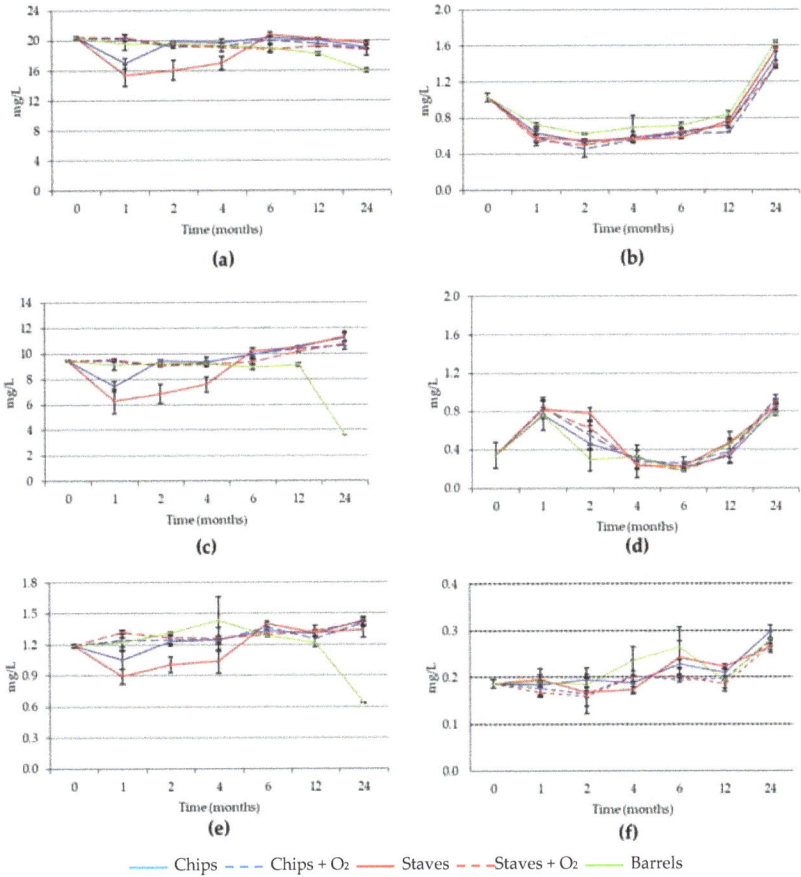

Figure 6. Evolution of cinnamic acids concentration (mg/L) over time: (**a**) caffeic acid; (**b**) caftaric acid; (**c**) coumaric acid; (**d**) coutaric acid; (**e**) ferulic acid; and (**f**) fertaric acid. Mean ± standard deviation ($n = 2$).

Coumaric acid concentration in wines may decrease in the presence of the enzyme cinnamate decarboxylase, by transformation into 4-vinylphenol, which can be transformed into 4-ethylphenol in the presence of a vinylphenol reductase. In the same way, ferulic acid can decrease by decarboxylation, transforming into 4-vinylguaiacol, and in the presence of the enzyme vinylphenol reductase can form 4-ethylguaiacol [45].

Caffeic, coumaric, and ferulic acids presented a more or less stable concentration up to 12 months, with a similar evolution for all treatments (Figure 6a,c,e). It is important to note the decrease of these acids at 12 months in wines aged in barrels, which could be explained by the decarboxylation reactions described in the previous paragraph. The decrease in coumaric and ferulic acids at 12 months coincided with a significant increase in 4-ethylphenol and 4-ethylguaiacol (Figure 4c,d), which as explained above, could be due to contamination of the wine by *Brettanomyces* at the time of bottling. On the other hand, Cadahía et al. [50] justified the variations of the concentration of caffeic acid due to their involvement in esterification reactions to give caftaric acid, as well as in the co-pigmentation reactions that allow the color of anthocyanins to stabilize.

There was no clear relationship between the content of the free cinnamic acids and their esterified forms. While the free form concentrations remained relatively constant or increased slightly, their corresponding esters had a heterogeneous evolution. Caftaric and coutaric acids diminished or maintained their concentration constantly during the first months, whereas the fertaric acid content remained practically stable. In these compounds there was a notable increase from 12 to 24 months (Figure 6b,d,f). Tartaric esters (caftaric and coutaric acids) decrease during aging because they are very reactive compounds and participate in oxidation processes. The only differences between treatments were observed in the caftaric acid, which had a slightly higher concentration in wines aged in oak barrels (Figure 6b), probably because this type of container encourages the esterification processes [41].

The evolution of flavanols and flavonols in wines can be seen in Figure 7. As can be observed, the concentrations of catechin and epicatechin decreased slightly until 12 months, producing a greater decrease in the case of the epicatechin from 12 to 24 months (Figure 7a,b). These results coincide with those obtained by other authors [18,22,41,50,51] who also obtained a decrease in these phenols over the aging period. This is related to their participation in the oxidative processes, polymerization, and condensation reactions with other compounds, favored in barrels by the continuous diffusion of oxygen [52]. On the other hand, Del Barrio-Galán et al. [53] found lower concentrations of flavanols (catechin and epicatechin) in wines and in model solutions treated with chips than in control wines, corroborating the theory that these compounds can be adsorbed on the surface of the wood. Modifications of the flavanols content between treatments during aging could be due to the reactivity of these compounds. Thus, their concentration may decrease due to oxidation and polymerization reactions and may increase due to the hydrolysis of higher oligomers [54].

Quercetin concentration was higher in wines without micro-oxygenation than in the micro-oxygenates and those aged in barrels, over the whole time studied (Figure 7c). The values of this compound decreased in wines aged in oak barrels and micro-oxygenated wines during the first two months, remaining practically constant during the rest of the process. Likewise, in wines without micro-oxygenation, its concentration remained more or less constant throughout the time studied. Cejudo-Bastante et al. [55] also found lower concentrations in wines aged with chips and micro-oxygenated, compared to those treated with chips but without micro-oxygenation. Castellari et al. [56] also observed a decrease of this compound in micro-oxygenated wines with respect to the control one.

Rutin concentration decreased significantly during the first month in all the wines, decreasing later in the micro-oxygenated wines and wines in barrels, and remaining practically constant in the non-micro-oxygenated wines during the rest of the time studied (Figure 7d). Fernández de Simón et al. [51] also observed a decrease in some flavonols in wines aged for 21 months in oak barrels of different origins. This decrease could be due to the fact that flavonols can react with anthocyanins in co-pigmentation reactions [57,58].

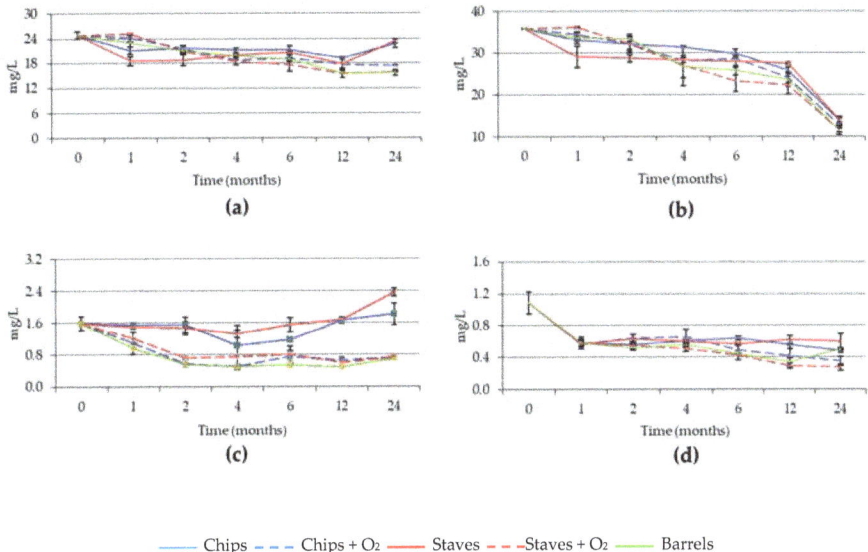

Figure 7. Evolution over time of flavanols concentration (mg/L): (**a**) catechin and (**b**) epicatechin; and of flavonols concentration (mg/L): (**c**) quercetin and (**d**) rutin. Mean ± standard deviation (*n* = 2).

Figure 8 shows the evolution of the concentration of *trans*-resveratrol and its glycoside, *trans*-piceid. The concentration of *trans*-piceid was hardly modified throughout the process, with no important differences being observed between the treatments (Figure 8a). Alañón et al. [46] did not find differences in the stilbene concentration between wines aged in chestnut wood barrels and wines in contact with chips.

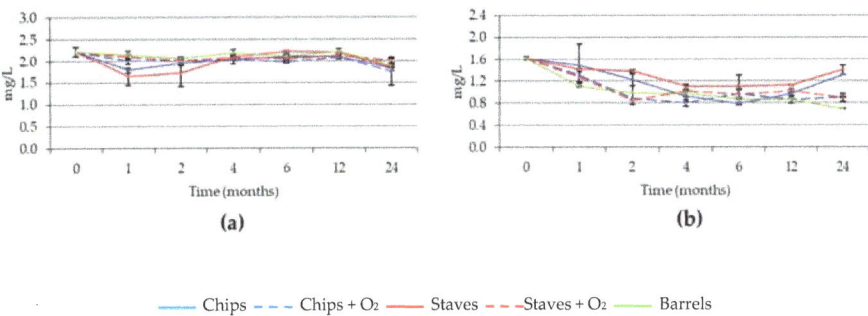

Figure 8. Evolution of stilbenes concentration (mg/L) over time: (**a**) *trans*-piceid; (**b**) *trans*-resveratrol. Mean ± standard deviation (*n* = 2).

Finally, the concentration of *trans*-resveratrol decreased in all treatments, more in micro-oxygenated wines and in those aged in oak barrels compared to non-micro-oxygenated wines (Figure 8b). Barrera-García et al. [59] estimated an average rate of decrease of *trans*-resveratrol of 50% in a model wine in the presence of oak wood, partly due to adsorption mechanisms on the surface of oak. This coincides with our results when considering micro-oxygenated wines and wines aged in barrels. The decline of stilbenes during aging was also observed by other authors [41,51,60].

3.4. Sensorial Analysis

Figure 9 shows the sensory evaluation of wines during the aging period considered. After 2 months, the most highly valued wines (lowest score) were those treated with staves, followed by those aged in barrel, while wines with chips scored worse. However, at 4, 6, and 12 months, the wines aged in barrels obtained better scores than those treated with fragments, and of these, the wines with staves were better scored than those with chips in most cases.

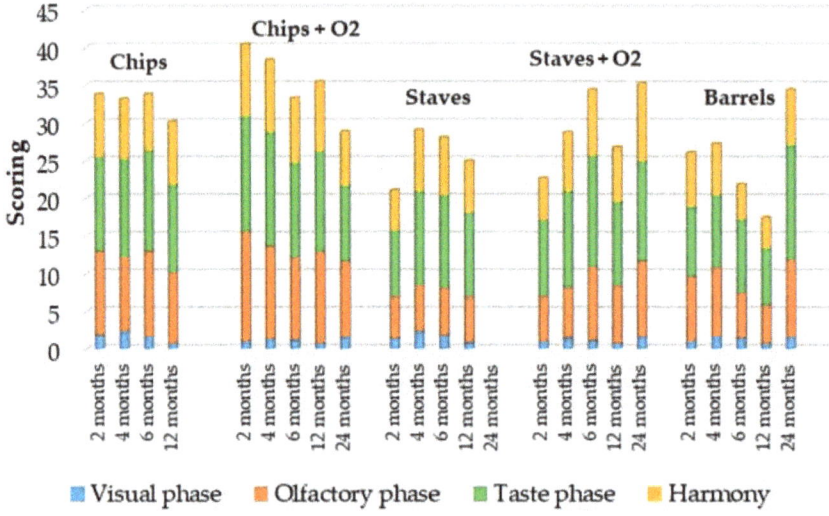

Figure 9. Sensory average valuation of wines over time.

Cano-López et al. [61] obtained the best sensory results in wines with fragments in the form of cubes compared to those aged in barrels, after 3 and 6 months of contact. For this reason, they affirmed that the use of oak chips could be a good choice for short aging wines.

At 24 months, only the sensory analysis of the micro-oxygenated wines and those aged in barrels was carried out. At this time the valuation changed, wines with chips obtaining the best score, followed by wines aged in barrels, and finally those treated with staves. The organoleptic deterioration of wine aged in barrels could be due to its contamination by *Brettanomyces* during bottling. In addition, in wines with staves a high volatile acidity was detected in the organoleptic analysis (data not shown).

Figure 10 shows the sensory profiles of the wines at the different moments studied. Fruity and varietal notes were hardly appreciated in the wines at any of the moments. Up to 6 months, notes related to wood were perceived more intensely in wines treated with staves (toasted, almond, caramel, vanilla, smoked), while at 12 months they were perceived equally in wines aged in barrels and treated with staves. This greater intensity of tertiary aromas in the wines with staves could possibly have been due to a more intense toasting of these, since these aromas come from compounds generated during the wood toasting. Casassa et al. [62] also found an increase in notes related to wood (toast, clove, vanilla, etc.) as the degree of toasting of the chips increased.

With respect to the gustatory phase, wines with chips were perceived as less structured, persistent, astringent, and with a lesser retronasal aroma after 2 months. However, at 4 months they were the ones considered best in this phase. Wines which were micro-oxygenated and in contact with staves obtained the best evaluations at 6 and 12 months, with scores at this time similar to those for wines aged in barrels.

Sensory evaluation of wines at 24 months could be considered invalid due to the strong impact that ethylphenols had on wines aged in barrels and the high volatile acidity in wines treated with staves, which did not allow an adequate evaluation of the rest of the attributes (data not shown).

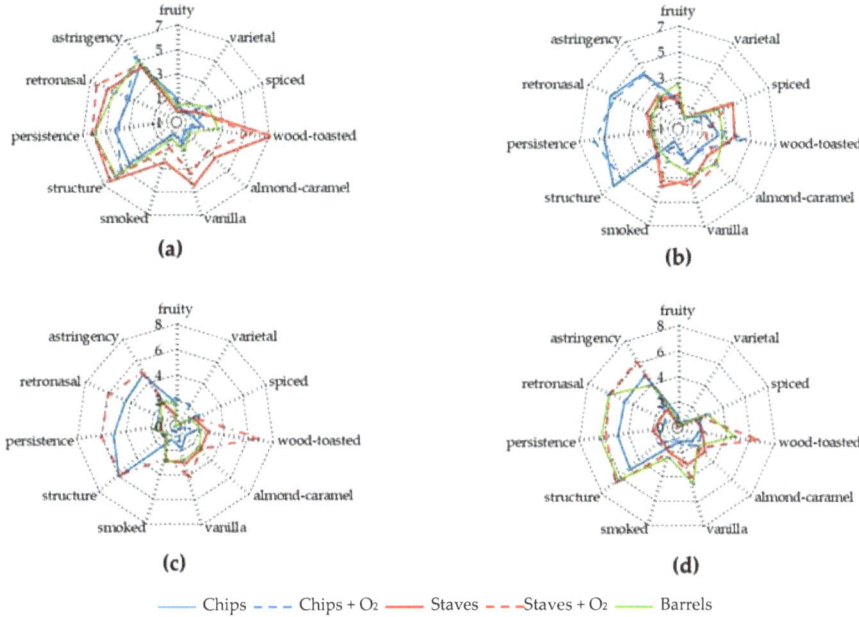

Figure 10. Sensory profiles of wines at: (**a**) 2 months; (**b**) 4 months; (**c**) 6 months; and (**d**) 12 months.

Regarding triangular tests, the results obtained in the sensory assessment at 6 and 12 months are shown in Table 1. At 6 months, wines added with fragments were clearly discriminated (with a level of significance of 99.9%) from those aged in barrels, independently of the size of the fragments. However, an adequate level of significance was not obtained for their discrimination at 12 months. When comparing the size of the fragments (chips vs. staves), it was possible to differentiate between the treatments at both moments (6 and 12 months) with a level of significance of 95% and 99% respectively.

As regards preferences, at 6 months, the preferred wines were those aged in barrels (72.2%), while at 12 months they did not opt for any of the two types of aging. In relation to the size of the fragments, the staves were preferred at 6 months (57.1%), and both treatments were evaluated equally at 12 months.

Table 1. Results of the sensory evaluation by triangular tests.

Moment	Variable	Mean Right Answer (Based on 10)	Preference (%)
6 months	Fragments-Barrel	10.0 (***)	72.2 barrel
	Chips-Staves	7.78 (*)	57.1 staves
12 months	Fragments-Barrel	6.43	50.0
	Chips-Staves	8.57 (**)	50.0

Significance level: $p < 0.05\%$ (*); $p < 0.01\%$ (**); $p < 0.001$ (***).

3.5. Treatments Classification

Two canonical discriminant analyses (DCA) were carried out to classify the different treatments based on the volatile compounds from oak wood and the low molecular weight phenols (Figure 11a,b, respectively). To carry out the discriminant analysis, the data from all the moments and all the trials studied were used in order to obtain as many variables as possible and try to achieve better sample classification. The graphs obtained present the distribution of the samples in the plane formed by the first two discriminant functions. When trying to discriminate the wines, a good separation was achieved between the three treatments (chips, staves, and barrels) for both groups of compounds.

In the case of the volatile composition (Figure 11a), discriminant function 1 explained 88.8% of the variance and discriminant function 2 explained 11.2%, representing 100% of all variance. In this graph, with 100% of cases correctly classified, the separation between the two types of aging was carried out by means of function 1. Wines aged in barrels were correlated with a high concentration of 5-hydroxymethylfurfural and 5-methylfurfural; and wines in contact with staves were correlated with a high content of vanillin and *trans*-isoeugenol, with these results coinciding with those shown in Section 3.2.

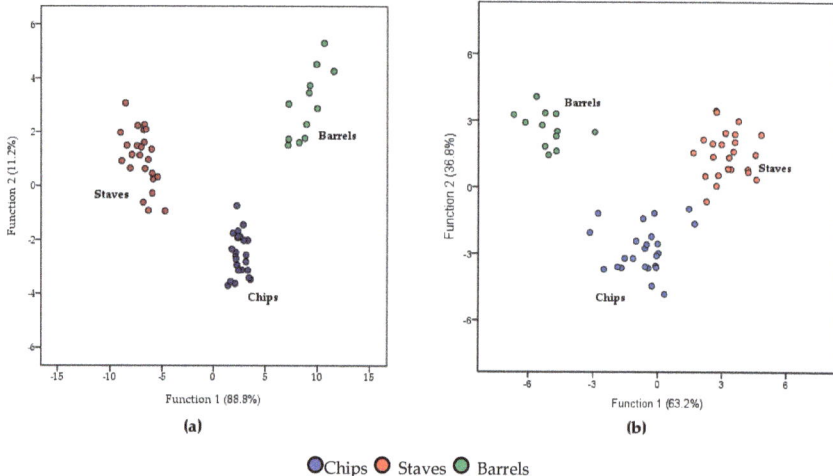

Figure 11. Canonical discriminant analysis: (**a**) volatile composition contributed by oak wood; and (**b**) low molecular weight phenols.

In the graph of low molecular weight polyphenols (Figure 11b), the percentage of accumulated variance explained by the first two functions was 100%, with discriminant function 1 explaining 63.2% of the variance, and discriminant function 2 the 36.8%. Additionally, 100% of the cases were correctly classified. Function 1 allowed us to differentiate between fragments and barrels, while function 2 separated the wines treated with chips from those aged in barrels and with staves. The variables with the highest discriminant power in function 1 were syringic acid and quercetin with a positive character (related to staves), and caftaric acid with a negative sign (associated with barrels). On the other hand, the discriminant function 2 correlates the caftaric acid with the wines aged in barrels and with fragments in the positive part of the axis, and the gallic acid with the wines treated with chips in the negative part. These results coincide with those obtained in Section 3.3.

4. Conclusions

The use of oak fragments and barrels during the aging of red wines influenced the evolution of the wines' chromatic parameters, which was similar between treatments over the aging time. Moreover, an improvement in volatile and phenolic compounds coming from wood was observed in all the wines. Due to the chromatic evolution of the wines and the contribution of substances which came from the wood, the optimal contact time between fragments and wine could be estimated as being 2 months.

The best sensory evaluation of the wines macerated with staves was obtained in short periods of aging, while for those aged in barrels, their sensory quality improved with time. Wines aged with chips were evaluated more poorly than the wines with other treatments at all times. Good sensory discrimination was also achieved between wines in contact with fragments and those aged in barrels.

The canonical discriminant analyses carried out with the volatile compounds and the low molecular weight phenols showed a good separation between the two types of aging (fragments and barrels), as well as between the two types of fragments (chips and staves).

The results of this study about the contact time with wood during the aging of red wines with oak wood fragments (chips and staves) indicated that the use of these fragments is a suitable practice for aging red wines, which can be an appropriate option for wines destined for short aging periods.

Author Contributions: Conceptualization and funding acquisition, J.M.; methodology, P.R.-B. and J.M.; data curation, P.R.-B. and J.M.; writing—original draft preparation, P.R.-B.; writing—review and editing, P.R.-B., T.G.-C. and J.M.; visualization and supervision, J.M. and T.G.-C.

Funding: This research was funded by Government of La Rioja for regional Projects: P.R.-12-06, P.R.-13-07, P.R.-11-08, P.R.-11-09, P.R.-11-10 and P.R.-15-11.

Acknowledgments: P.R.-B. would like to thank her predoctoral grant of Government of La Rioja. We would like to thank the wine professionals for their participation in sensory evaluation.

Conflicts of Interest: The authors declare no conflict of interest.

References

1. Ribéreau-Gayón, P.; Glories, Y.; Maujean, A.; Dubourdieu, D. *Tratado de Enología 2. Química del Vino. Estabilización y Tratamientos*; Ediciones Mundi-Prensa: Madrid, Spain, 2003; ISBN 978-9-50504-571.

2. Rubio-Bretón, P.; Lorenzo, C.; Salinas, M.R.; Martínez, J.; Garde-Cerdán, T. Influence of Oak Barrel Aging on the Quality of Red Wines. In *Oak: Ecology, Types and Management*; Chuteira, C.A., Grao, A.B., Eds.; Nova Science Publishers: New York, NY, USA, 2013; Volume 2, pp. 59–86, ISBN 978-1-61942-493-7.

3. Resolution OENO 9/2001. Usage of pieces of oak wood in winemaking. In *International Codex of Oenological Practices*; Office International de la Vigne et du Vin: Paris, France, 2001.

4. Resolution OENO 3/2005. Pieces of oak wood. In *International Codex of Oenological Practices*; Office International de la Vigne et du Vin: Paris, France, 2005.

5. European Commission (EC). Council regulation (EC) No. 2165/2005 of 20 December 2005 amending regulation (EC) No. 1493/1999 on the common organisation of the market in wine. *Off. J. Eur. Communities* **2005**, *L345*, 1–4.

6. European Commission (EC). Commission regulation (EC) No. 1507/2006 of 11 October 2006 amending regulations (EC) No. 1622/2000, (EC) No. 884/2001 and (EC) No. 753/2002 concerning certain detailed rules implementing Regulation (EC) No. 1493/1999 on the common organisation of the market in wine, as regards the use of pieces of oak wood in winemaking and the designation and presentation of wine so treated. *Off. J. Eur. Communities* **2006**, *L280*, 9–11.

7. European Commission (EC). Commission regulation (EC) No. 606/2009 of 10 July 2009 laying down certain detailed rules for implementing Council Regulation (EC) No. 479/2008 as regards the categories of grapevine products, oenological practices and the applicable restrictions. *Off. J. Eur. Communities* **2009**, *L193*, 1–59.

8. Chatonnet, P. Productos alternativos a la crianza en barrica de los vinos. Influencia de los parámetros de fabricación y de uso. *Rev. Enol.* **2007**, *4*, 2–24.

9. Verdier, B.; Blateyron, L.; Granès, D. Las virutas y los bloques: Como razonar sobre su puesta en práctica. In *Crianza en Barricas y Otras Alternativas*; Fundación para la Cultura del Vino: Madrid, Spain, 2007; Volume 2, pp. 191–196.

10. European Commission (EC). Commission regulation (EC) No. 2676/1990 of 17 September 1990 determining Community methods for the analysis of wine. *Off. J. Eur. Communities* **1990**, *L272*, 1–192.

11. Glories, Y. La couleur des vins rouges. 2ème partie. Mesure, origine et interprétation. *Connais. Vigne Vin* **1984**, *18*, 253–271. [CrossRef]

12. Ribéreau-Gayon, P.; Stonestreet, E. Determination of anthocyanins in red wine. *Bull. Soc. Chim. Fr.* **1965**, *9*, 2649–2652. [PubMed]

13. Ortega, C.; López, R.; Cacho, J.; Ferreira, V. Fast analysis of important wine volatile compounds. Development and validation of a new method based on gas chromatographic-flame ionisation detection analysis of dichloromethane microextracts. *J. Chromatogr. A* **2001**, *923*, 205–214. [CrossRef]

14. Martínez, J.; Rubio-Bretón, P. Composición fenólica no antociánica en vinos de Tempranillo de la D.O.Ca. Rioja. Efecto de la añada, el empleo de microoxigenación y la crianza en barrica. In *Zubía. Revista de Ciencias. Monográfico Núm. 25*; Instituto de Estudios Riojanos: Logroño, Spain, 2013; Volume 4, pp. 31–45, ISSN 1131-5423.

15. AENOR. *Análisis Sensorial. Recopilación de Normas UNE*; AENOR: Madrid, Spain, 1997; ISBN 978-84-8143-705-8.

16. Laqui-Estaña, J.; López-Solís, R.; Peña-Neira, A.; Medel-Maraboli, M.; Obreque-Slier, E. Wines in contact with oak wood: The impact of the variety (Carménère and Cabernet Sauvignon), format (barrels, chips and staves), and aging time on the phenolic composition. *J. Sci. Food Agric.* **2018**. [CrossRef] [PubMed]

17. Martínez-Gil, A.M.; del Álamo-Sanza, M.; Gutiérrez-Gamboa, G.; Moreno-Simunovic, Y.; Nevares, I. Volatile composition and sensory characteristics of Carménère wines macerating with Colombian (*Quercus humboldtii*) oak chips compared to wines macerated with American (*Q. alba*) and European (*Q. petraea*) oak chips. *Food Chem.* **2018**, *266*, 90–100. [CrossRef] [PubMed]

18. Cano-López, M.; Pardo-Mínguez, F.; López-Roca, J.M.; Gómez-Plaza, E. Effect of microoxygenation on anthocyanin and derived pigment content and chromatic characteristics of red wines. *Am. J. Enol. Vitic.* **2006**, *57*, 325–331.

19. Oberholster, A.; Elmendorf, B.L.; Lerno, L.A.; King, E.S.; Heymann, H.; Brenneman, C.E.; Boulton, R.B. Barrel maduration, oak alternatives and micro-oxygenation: Influence on red wine aging and quality. *Food Chem.* **2015**, *173*, 1250–1258. [CrossRef] [PubMed]

20. Del Álamo, M.; Nevares, I.; Merino-García, S. Influence of different aging systems and oak woods on aged wine color and anthocyanin composition. *Eur. Food Res. Technol.* **2004**, *219*, 124–132. [CrossRef]

21. Pérez-Prieto, L.J.; De la Hera-Orts, M.L.; López-Roca, J.M.; Fernández-Fernández, J.I.; Gómez-Plaza, E. Oak-matured wines: Influence of the characteristics of the barrel on wine colour and sensory characteristics. *J. Sci. Food Agric.* **2003**, *83*, 1445–1450. [CrossRef]

22. Tavares, M.; Jordao, A.M.; Ricardo-da-Silva, J.M. Impact of cherry, acacia and oak chips on red wine phenolic parameters and sensory profile. *OENO One* **2017**, *51*, 329–342. [CrossRef]

23. Cadahía, E.; Fernández de Simón, B.; Sanz, M.; Poveda, P.; Colio, J. Chemical and chromatic characteristics of Tempranillo, Cabernet Sauvignon and Merlot wines from D.O. Navarra aged in Spanish and French oak barrels. *Food Chem.* **2009**, *115*, 639–649. [CrossRef]

24. Garde-Cerdán, T.; Torrea-Goñi, D.; Ancín-Azpilicueta, C. Accumulation of volatile compounds during ageing of two red wines with different composition. *J. Food Eng.* **2004**, *65*, 349–356. [CrossRef]

25. Towey, J.P.; Waterhouse, A.L. The extraction of volatile compounds from French and American oak barrels in Chardonnay during three successive vintages. *Am. J. Enol. Vitic.* **1996**, *47*, 163–172.

26. Garde-Cerdán, T.; Ancín-Azpilicueta, C. Effect of oak barrel type on the volatile composition of wine: Storage time optimization. *LWT-Food Sci. Technol.* **2006**, *39*, 199–205. [CrossRef]

27. Pérez-Prieto, L.J.; López-Roca, J.M.; Martínez-Cutillas, A.; Pardo-Mínguez, F.; Gómez-Plaza, E. Extraction and formation dynamic of oak-related volatile compounds from different volume barrels to wine and their behavior during bottle storage. *J. Agric. Food Chem.* **2003**, *51*, 5444–5449. [CrossRef]

28. Boidron, J.N.; Chatonnet, P.; Pons, M. Influence du bois sur certaines substances odorantes des vins. *Connais. Vigne Vin* **1988**, *22*, 275–294. [CrossRef]

29. Tominaga, T.; Blanchard, L.; Darriet, P.; Dubourdieu, D. A powerful aromatic volatile thiol, 2-furanmethanethiol, exhibiting roast coffee aroma in wines made from several *Vitis vinifera* grape varieties. *J. Agric. Food Chem.* **2000**, *48*, 1799–1802. [CrossRef] [PubMed]

30. Nonier-Bourden, M.F.; Vivas, N.; Absalon, C.; Vitry, C.; Fouquet, E.; Vivas de Gaulejac, N. Structural diversity of nucleophilic adducts from flavanols and oak wood aldehydes. *Food Chem.* **2008**, *107*, 1494–1505. [CrossRef]

31. Garde-Cerdán, T.; Ancín-Azpilicueta, C. Review of quality factors on wine ageing in oak barrels. *Trends Food Sci. Tech.* **2006**, *17*, 438–447. [CrossRef]

32. Jarauta, I.; Cacho, J.; Ferreira, V. Concurrent phenomena contributing to the formation of the aroma of wine during aging in oak wood: An analytical study. *J. Agric. Food Chem.* **2005**, *53*, 4166–4177. [CrossRef] [PubMed]

33. Fernández de Simón, B.; Cadahía, E.; Del Álamo, M.; Nevares, I. Effect of size, seasoning and toasting in the volatile compounds in toasted oak wood and in a red wine treated with them. *Anal. Chim. Acta* **2010**, *660*, 211–220. [CrossRef] [PubMed]

34. Spillman, P.J.; Pollnitz, A.P.; Liacopoulos, D.; Pardon, K.H.; Sefton, M.A. Formation and degradation of furfuryl alcohol, 5-methylfurfuryl alcohol, vanillyl alcohol, and their ethyl ethers in barrel aged wines. *J. Agric. Food Chem.* **1998**, *46*, 657–663. [CrossRef]

35. Ferreira, V.; Jarauta, I.; Cacho, J. Physicochemical model to interpret the kinetics of aroma extraction during wine aging in wood. Model limitations suggest the necessary existence of biochemical processes. *J. Agric. Food Chem.* **2006**, *54*, 3047–3054. [CrossRef]

36. Spillman, P.J.; Pollnitz, A.P.; Liacopoulos, D.; Skouroumounis, G.K.; Sefton, M.A. Accumulation of vanillin during barrel-aging of white, red, and model wines. *J. Agric. Food Chem.* **1997**, *45*, 2584–2589. [CrossRef]

37. Spillman, P.J.; Iland, P.G.; Sefton, M.A. Accumulation of volatile oak compounds in a model wine stored in American and Limousin oak barrels. *Aust. J. Grape Wine R.* **1998**, *4*, 67–73. [CrossRef]

38. Garde-Cerdán, T.; Goñi-Torrea, D.; Ancín-Azpilicueta, C. Changes in the concentration of volatile oak compounds and esters in red wine stored for 18 months in re-used French oak barrels. *Aust. J. Grape Wine Res.* **2002**, *8*, 140–145. [CrossRef]

39. Fernández de Simón, B.; Cadahía, E.; Jalocha, J. Volatile compounds in a Spanish red wine aged in barrels made of Spanish, French, and American oak wood. *J. Agric. Food Chem.* **2003**, *51*, 7671–7678. [CrossRef] [PubMed]

40. Garde-Cerdán, T.; Lorenzo, C.; Carot, J.M.; Esteve, M.D.; Climent, M.D.; Salinas, M.R. Effects of composition, storage time, geographic origin and oak type on the accumulation of some volatile oak compounds and ethylphenols in wines. *Food Chem.* **2010**, *122*, 1076–1082. [CrossRef]

41. Martínez, J. Incidencia del Origen de la Madera de Roble en la Calidad de los Vinos de Tempranillo de la D.O.Ca. Rioja Durante la Crianza en Barrica. Doctoral's Thesis, Universidad de La Rioja, La Rioja, Spain, 2004.

42. Waterhouse, A.L.; Towey, J.P. Oak lactone isomer ratio distinguishes between wines fermented in American and French oak barrels. *J. Agric. Food Chem.* **1994**, *42*, 1971–1974. [CrossRef]

43. Pérez-Prieto, L.J.; López-Roca, J.M.; Martínez-Cutillas, A.; Pardo-Mínguez, F.; Gómez-Plaza, E. Maturing wines in oak barrels. Effects of origin, volume, and age of the barrel on the wine volatile composition. *J. Agric. Food Chem.* **2002**, *50*, 3272–3276. [CrossRef] [PubMed]

44. Suárez, R.; Suárez-Lepe, J.A.; Morata, A.; Calderón, F. The production of ethylphenols in wine by yeasts of the genera Brettanomyces and Dekkera: A review. *Food Chem.* **2007**, *102*, 10–21. [CrossRef]

45. Chatonnet, P.; Dubourdieu, D.; Boidron, J.N.; Pons, M. The origin of ethylphenols in wines. *J. Sci. Food Agric.* **1992**, *60*, 165–178. [CrossRef]

46. Alañón, M.E.; Schumacher, R.; Castro-Vázquez, L.; Díaz-Maroto, M.C.; Hermosín-Gutiérrez, I.; Pérez-Coello, M.S. Enological potential of chestnut wood for aging Tempranillo wines. Part II: Phenolic compounds and chromatic characteristics. *Food Res. Int.* **2013**, *51*, 536–543. [CrossRef]

47. Del Álamo, M.; Nevares, I.; Cárcel-Cárcel, L.M.; Navas-Gracia, L. Analysis for low molecular weight phenolic compounds in a red wine aged in oak chips. *Anal. Chim. Acta* **2004**, *513*, 229–237. [CrossRef]

48. Sroka, Z.; Cisowski, W. Hydrogen peroxide scavenging, antioxidant and anti-radical activity of some phenolic acids. *Food Chem. Toxicol.* **2003**, *41*, 753–758. [CrossRef]

49. Sanz, M.; Cadahía, E.; Esteruelas, E.; Muñoz, A.M.; Fernández de Simón, B.; Hernández, T.; Estrella, I. Phenolic compounds in chestnut (*Castanea sativa* Mill.) heartwood. Effect of toasting at cooperage. *J. Agric. Food Chem.* **2010**, *58*, 9631–9640. [CrossRef] [PubMed]

50. Cadahía, E.; Fernández de Simón, B.; Poveda, P.; Sanz, M. *Utilización de Quercus pyrenaica Willd. de Castilla y León en el Envejecimiento de vinos: Comparación con Roble Francés y Americano*; Instituto Nacional de Investigación y Tecnología Agraria y Alimentaria, Ministerio de Ciencia e Innovación: Madrid, Spain, 2008; ISBN 978-8-47498-525-2.

51. Fernández de Simón, B.; Hernández, T.; Cadahía, E.; Dueñas, M.; Estrella, I. Phenolic compounds in a Spanish red wine aged in barrels made of Spanish, French and American oak wood. *Eur. Food Res. Technol.* **2003**, *216*, 150–156. [CrossRef]

52. Laszlavik, M.; Gál, L.; Misik, S.; Erdei, L. Phenolic compounds in two Hungarian red wines matured in *Quercus robur* and *Quercus petraea* barrels: HPLC analysis and Diode Array Detection. *Am. J. Enol. Vitic.* **1995**, *46*, 67–74.

53. Del Barrio-Galán, R.; Ortega-Heras, M.; Sánchez-Iglesias, M.; Pérez-Magariño, S. Interactions of phenolic and volatile compounds with yeast lees, commercial yeast derivatives and non-toasted chips in model solutions and young red wines. *Eur. Food Res. Technol.* **2012**, *234*, 231–244. [CrossRef]

54. Dallas, C.; Ricardo-Da Silva, J.M.; Laureano, O. Degradation of oligomeric procyanidins and anthocyanins in a Tinta Roriz red wine during maturation. *Vitis* **1995**, *34*, 51–56.

55. Cejudo-Bastante, M.J.; Hermosín-Gutiérrez, I.; Pérez-Coello, M.S. Micro-oxygenation and oak chip treatments of red wines: Effects on colour-related phenolics, volatile composition and sensory characteristics. Part II: Merlot wines. *Food Chem.* **2011**, *124*, 738–748. [CrossRef]

56. Castellari, M.; Matricardi, L.; Arfelli, G.; Galassi, S.; Amati, A. Level of single bioactive phenolics in red wine as a function of the oxygen supplied during storage. *Food Chem.* **2000**, *69*, 61–67. [CrossRef]

57. Boulton, R. The copigmentation of anthocyanins and its role in the color of red wine: A critical review. *Am. J. Enol. Vitic.* **2001**, *52*, 67–87.

58. Schwarz, M.; Picazo-Bacete, J.J.; Winterhalter, P.; Hermosín-Gutiérrez, I. Effect of copigments and grape cultivar on the color of red wines fermented after the addition of copigments. *J. Agric. Food Chem.* **2005**, *53*, 8372–8381. [CrossRef]

59. Barrera-García, V.D.; Gougeon, R.D.; Di Majo, D.; De Aguirre, C.; Voilley, A.; Chassagne, D. Different sorption behaviors for wine polyphenols in contact with oak wood. *J. Agric. Food Chem.* **2007**, *55*, 7021–7027. [CrossRef]

60. Hernández, T.; Estrella, I.; Carlavilla, D.; Martín-Álvarez, P.J.; Moreno-Arribas, M.V. Phenolic compounds in red wine subjected to industrial malolactic fermentation and ageing on lees. *Anal. Chim. Acta* **2006**, *563*, 116–125. [CrossRef]

61. Cano-López, M.; Bautista-Ortín, A.B.; Pardo-Mínguez, F.; López-Roca, J.M.; Gómez-Plaza, E. Sensory descriptive analysis of a red wine aged with oak chips in stainless steel tanks or used barrels: Effect of the contact time and size of the oak chips. *J. Food Qual.* **2008**, *31*, 645–660. [CrossRef]

62. Casassa, F.; Sari, S.; Avagnina, S.; Catania, C. Efecto del empleo de chips de roble y del tipo de tostado sobre la composición polifenólica y las características cromáticas y organolépticas de vinos cv. Merlot. *Vitic./Enol. Prof.* **2008**, *116*, 22–35.

Review

New Strategies to Improve Sensorial Quality of White Wines by Wood Contact

M. Elena Alañón [1,*] , M. Consuelo Díaz-Maroto [1] and M. Soledad Pérez-Coello [2]

1   Area of Food Technology, Regional Institute for Applied Scientific Research (IRICA), University of Castilla-La Mancha, Avda. Camilo José Cela, 10, 13071 Ciudad Real, Spain; mariaconsuelo.diaz@uclm.es

2   Area of Food Technology, Faculty of Chemical Sciences and Technologies, University of Castilla-La Mancha, Avda. Camilo José Cela, 10, 13071 Ciudad Real, Spain; Soledad.perez@uclm.es

*   Correspondence: mariaelena.alanon@uclm.es; Tel.: +34-926295300; Fax: +34-926295318

Received: 31 October 2018; Accepted: 16 November 2018; Published: 21 November 2018

Abstract: Ageing wine is a common practice used in winemaking, since the quality and sensory profile increase due to the extractable compounds coming from wood, by means of barrels or chips. The quantitative and qualitative compounds of the wood depend on the species, its origins and the treatments applied in cooperages. Traditionally, oak wood species are most often used in cooperage, specifically *Quercus alba (Q. alba)*, Known as American oak and *Quercus robur (Q. robur)* and *Quercus petraea (Q. petraea)*, both known as French oak. Although this stage is very common for red wines, its use is still restricted in the case of white wines. However, this topic is particularly interesting, since due to the sensorial benefits of wood contact, the option for ageing white wines in barrels or chips could be chosen by winemakers. This review compiles the novel strategies applied to white wines by means of wood contact in recent years with the aim to increase wine quality and sensorial features.

Keywords: white wine; volatile compounds; sensorial characteristics; oak; alternative woods; barrels; chips

---

1. Introduction

The white wine market has been monopolized for many years by young varietal wines, which should be consumed in a short period after bottling to avoid the loss of freshness and fruity character, mainly due to the detriment of the compounds of varietal origin. Different oenological practices were used to enhance varietal aromas in slightly aromatic or neutral white grape varieties, such as pre-fermentative skin-contact treatment or the use of glycosidic enzymes, to obtain more aromatic wines with a low content of phenolic compounds [1–3].

Wood ageing in the case of white wines has only been used occasionally, mainly by fermentation in barrels or by ageing on lees [4,5]. The yeast cells absorb the ellagitannins from the wood, reducing the astringency of the wines, while some compounds extracted from wood such as furfural and vanillin can be metabolized by the yeasts decreasing their sensory impact [6].

Recently, these techniques have been recovered together with other alternative practices in white vinification, such as the use of chips which are allowed in some countries. Fermentation and/or ageing in wooden barrels, accelerated ageing with wood chips, ageing in barrique on lees [5,7–9] and the use of other types of wood such as acacia or cherry [10,11] have been used to improve the quality of white wines.

Different authors have studied the volatile composition of different woods for potential use in enology [12–14]. Table 1 shows the concentration of some of the main volatile compounds found in those "new" woods, such as chestnut, acacia or cherry, along with the components of the oak woods traditionally used in the aging of wines.

All these innovations have contributed to increasing and improving the variety of white wines in contact with wood in the market, adapting to the new tastes of the consumer and the demands of the international market. However, it is important to look for the best combination between type of wood, ageing process and grape variety in order to obtain a quality wine with new sensory sensations, but without masking the primary and secondary aromas specific to each grape variety.

Table 1. Volatile composition (µg/g) of different untoasted woods with potential use in enology [13,14].

Compounds	American Oak (Q. alba)	French Oak (Q. petraea)	French Oak (Q. robur)	Chestnut	Acacia	Cherry
Furfural	5.79	12.09	17.90	3.55	0.56	-
5-Methylfurfural	0.41	3.57	4.95	2.34	0.05	-
3-Oxo-α-ionol	0.03	0.55	1.10	-	-	9.17
trans-Oak lactone	1.64	2.14	2.87	-	-	-
cis-Oak lactone	39.37	6.12	3.19	-	-	-
Guaiacol	0.91	4.46	4.22	5.24	0.10	0.16
Methylguaiacol	0.24	1.11	0.92	1.51	-	0.01
p-Vinylguaiacol	3.37	3.42	4.58	2.99	0.37	0.65
Syringol	0.41	3.42	3.18	3.18	0.16	0.10
Eugenol	3.44	1.05	1.39	2.53	0.92	0.11
Vanillin	70.37	45.69	6.40	80.90	4.70	4.68
Isoeugenol	0.36	0.30	0.15	1.24	1.19	1.31
Acetovanillone	2.73	3.41	0.47	4.77	0.14	0.17
Propiovanillone	1.92	1.64	1.11	0.62	-	1.37
Butyrovanillone	7.40	6.86	2.15	10.16	0.47	1.32
4-Allylsyringol	0.92	0.89	0.63	1.46	0.45	1.13
Syringaldehyde	23.21	26.13	24.38	21.90	12.23	-
Ethyl vanillate	1.88	6.54	5.29	2.68	1.16	0.20
4-Propenylsyringol	1.60	3.10	1.25	-	0.87	0.99
Coniferaldehyde	23.11	29.05	25.87	29.20	9.76	2.72
Sinapaldehyde	4.34	7.67	6.30	5.24	10.56	3.56

2. Sensorial Quality Improvement of White Wines by Contact with Oak Wood

In recent years, oenological research has not only focused on the ageing of red wines. There are more and more studies on the contact between white wine and wood, either during the ageing stage and/or during alcoholic fermentation. In this sense, the selection of wood type is fundamental. The barrels traditionally used are those of American oak (*Quercus alba*) and French oak (*Q. robur* and *Q. petraea*), the latter having the most prestige in the market and, therefore, also higher price. Both species are not exclusive to France, but extend across Eastern Europe and the North of the Iberian Peninsula, so lately Hungarian, Russian or Spanish oak barrels have been introduced into the market. These new oak woods present a similar volatile composition to that of French ones, and they are a good alternative in the ageing of wines [15,16].

The knowledge of the chemical composition of oak wood, especially its content in volatile compounds and ellagitannins, is of great importance to select the most suitable oak for the ageing of a particular wine. During ageing time, the transfer of several chemical substances takes place from the wood to the wine, which will condition the sensory characteristics of the final product. In different studies, it has been shown that there are very few chemical characteristics of a single oak species. On the contrary, there is a great variability within each species and even within each geographical area [15,17], since the climate and forest conditions have a great influence on chemical composition of oak wood [18]. Moreover, the physical and mechanical parameters of the wood (porosity, grain size, flexibility, etc.), can have an important influence on the transfer of substances between wine and wood, for that reason, the most appreciated species are those commonly known as "porous ring" woods, such as the oak.

To provide the woody character to white wines, fermentation or ageing in oak barrels have been carried out. However, compared to red wines, there are few studies on the effect of the interaction between oak wood barrels and white wine. Although, one of the most suitable white grape varieties to ferment the must or to age the wine in oak barrels is Chardonnay [5,19–22]. In the literature, different studies about the influence of oak maturation on the quality of white wines from other varieties can be found, such as Verdejo [23], white Listán [24], Muscatel [25], Sauvignon blanc [20,22], Encruzado [26] or Malvazija istarska [9].

Herrero et al. [22] studied the effect of toasting level and ageing time on the volatile composition and sensory quality of two white wines, Chardonnay and Sauvignon blanc, aged in oak barrels. The differences observed in the content in wood-extractable volatile compounds of wines were mainly dependent on the toasting level and ageing time. The volatile compounds released by the oak wood into wines increased with the ageing time, except methyl vanillate and vinylphenols, which decreased, although the authors found a high variability among replicates, which has been previously reported in other works [21]. From a sensory point of view, Chardonnay and Sauvignon blanc wines 12-month aged in French barrels did not show an homogeneous aroma quality among judges [22]. Previous studies that had evaluated the impact of ageing in oak barrels in Chardonnay wines are by Spillman et al. [19] and Herjavec et al. [20], among others. The former established correlations between the volatile composition and the olfactory profile of wines aged in new oak barrels and confirmed the role of some oak wood-extractable compounds in aged Chardonnay wine aroma, such as the strong correlation between "smoky" aroma and volatile components produced by barrel toasting. While Herjavec et al. [20] observed a positive influence of the fermentation in new Croatian oak barrels on the sensory properties of Chardonnay and Sauvignon blanc wines, in comparison with those fermented in steel tanks.

The aroma quality of Chardonnay wines has also been tried to be improved by fermentation and ageing in oak barrels on lees. This technology allowed increasing the content of volatile compounds positively related to the quality of the wines' flavour. However, no significant differences were observed in the aroma of the wines compared to the fermented and aged wines in stainless steel, probably due to the stronger influence of lees in the aroma of the wines compared to oak wood [5].

On the other hand, Rodríguez-Nogales et al. [23] classified Verdejo wines obtained through different winemaking techniques based on their volatile composition and sensory characteristics. Verdejo wines fermented and aged in oak barrels shown greater amounts of eugenol and methyleugenol, and lower quantities of terpenes and esters compared to those young wines fermented in stainless steel tanks. More recently, Lukic et al. [9] increased the aromatic composition complexity of white wines by prolonged maceration followed by maturation in wooden barrels. The utilization of new oak wood barrels induced changes for some aroma descriptors, such as "wood aroma" and "aroma intensity" in Encruzado white wines [26]. The authors showed the important role of ageing time and barrel capacity on the quality of oak barrel-aged white wines. The major impact on the evolution of sensorial properties of Encruzado wines was observed in new oak barrels of 225 L during 180 aging days.

However, these oenological practices have some drawbacks such as the limited capacity of the barrels, mainly when they are used for wine fermentation, the difficulty to control the temperature of fermentation or the cleaning of the barrels. Also, not all wines are suitable for ageing in oak barrels since the micro-diffusion of oxygen through wood pores could oxidize the wine, and the release of chemicals into the wine could completely mask its sensorial characteristics [27]. In the case of white wines, oxidation causes a decrease in pleasant sensory levels along with the appearance of off-flavours as "honey-like" or "cooked vegetables", and the brown coloration of wine [28]. In this sense, oak wood chips are presented as a good alternative to the use of barrels for the ageing of white wines. Their price is less than that of barrels, however, the different behaviour of the chips has made the evolution of this technique slower than it might seem.

Like barrels, chips can be used during fermentation or during wine ageing. Various authors have revealed the advantages of the fermentation of white wines in the presence of oak chips [24,29,30]. Airén wines were fermented with untreated chips from the most common oak varieties, American and French, at different doses, 4, 7 and 14 g/L. Wines best valued by the tasters were those treated with 7g/L of American oak chips [29].

On the other hand, Gutiérrez Alfonso [24] compared from the sensorial point of view two *Listán blanco* white wines fermented with oak chips and in barrels. Chips from American and French oaks were added in two doses, 4 and 8 g/L, while new barrels of 225 L were made from the same type of oak. The variable with the greatest effect on the sensory profile of the wine was the amount of oak chips used, while the geographical origin of the oak was more noticeable in chips than in barrels. In this sense, wines fermented with the higher quantities of American oak chips showed the greater intensities of vanilla and coconut aromas, together a greater astringency than in barrels.

Recently, Sanchéz-Palomo et al. [30] tried to improve the quality of Verdejo white wines by using medium-toasted oak chips during the alcoholic fermentation of the must or during the ageing of the wine. Wines aged in the presence of chips showed the highest quantities of volatile oak-extractable compounds, such as oak lactones and furanic compounds, while wines fermented with chips had higher concentrations of fermentative volatile substances, as alcohols, acetates and ethyl esters of straight-chain fatty acids.

In respect to the use of oak chips during ageing, the volatile composition and the sensory characteristics of Chardonnay wines treated with chips of different oak species and toasting degree have been studied [7]. Table 2 shows the odour activity values (OAV) of volatile oak-related compounds in Chardonnay wines aged with non-toasted and toasted American (*Q. alba*) and Hungarian (*Q. petraea*) oak chips during 25 days, together the odour perception thresholds and odour descriptions of each compound are found in the literature [16,31–37]. Volatile oak-related compounds with higher OAVs were those quantified in wines treated with American oak chips. Compounds with OAVs > 1 and with a possible impact on the aroma of the aged Chardonnay wine were cis-oak lactone, eugenol and 4-vinylguaiacol in wines treated with non-toasted American oak chips, and cis-oak lactone, eugenol, isoeugenol, guaiacol and vanillin in wines treated with toasted American oak chips. These results show the important effect that the toasting process has on the aromatic potential of oak, exhibiting those toasted oak samples higher number of compounds with OAVs > 1.

Table 2. Odour perception thresholds. Odour descriptions and odour activity values (OAV) of volatile oak-related compounds in Chardonnay wines macerated with non-toasted and toasted Hungarian (*Q. petraea*) and American (*Q. alba*) oak chips during 25 days.

Compounds	Odour Perception Threshold (µg/L)	Odour Description	OAV			
			NTA	TA	NTH	TH
trans-Oak lactone	122[b]	Vanilla, oaky, clove, coconut[a]	0.36	0.14	0.00 *	0.00
cis-Oak lactone	35[b]	Vanilla, oaky, clove, coconut[a]	9.11	3.17	0.00	0.00
Eugenol	5[b]	Spicy, clove, cinnamon[a]	2.76	1.24	0.58	0.60
Isoeugenol	6[e]	Spicy, clove, woody/oak[a]	0.37	1.98	0.45	1.32
Guaiacol	15[d]	Spicy, toasty, smoky/burnt[a]	0.29	1.33	0.13	2.63
4-Vinylguaiacol	141[d]	Smoky[d]	1.08	0.50	0.99	0.96
4-Ethylguaiacol	46[b]	Toasted bread, smoky, clove[b]	-	0.09	0.01	0.13
Syringol	570[f]	Smoky[c]	0.02	0.12	0.01	0.16
Vanillin	60[g]	Sweet, vanilla[a]	0.16	1.48	0.16	2.26
Furfural	15,000[b]	Slightly toasty, caramel[a]	0.00	0.01	0.00	0.03
5-Methylfurfural	16,000[b]	Spicy, toasty, sweet[a]	-	-	0.00	0.00
Maltol	5000[h]	Caramel[h]	-	0.05	-	0.01

NTA: Non-toasted American; TA: toasted American; NTH: non-toasted Hungarian; TH: toasted Hungarian. * Those compounds with OAV equal to zero were detected in trace amounts in wines and could not be quantified. [a] [16], [b] [31], [c] [2], [d] [33], [e] [34], [f] [35], [g] [36], [h] [37].

3. Sensorial Quality Improvement of White Wines by Contact with Alternative Woods to Oak

Although oak is by far the wood most used to carry out the ageing process, in recent years several studies have been published relating to the effects of other woods such as chestnut, cherry and acacia in wine ageing [38–47]. However, the scientific researchers have been focused on the impact of the use of alternative wood species exclusively on the quality of red wines and not of white wines. Therefore, there is a restricted knowledge about the impact of alternative woods to oak wood on white wine quality during the ageing process.

Recently, the effect of chips from different types of woods such as acacia and cherry on the ageing process of Encruzado wines was evaluated in comparison with the traditional chips from American and French oak woods [11]. As consequence of the wood contact, all aged white aged wines had increased polyphenol content due to phenols transfer from wood to wine. However, those wines aged with acacia chips exhibited the highest total phenolic content (342.94 mg/L) in comparison with the rest of wines aged with cherry and oak wood (328.59 mg/L and 319.86 mg/L, respectively) [11]. The particular richness of some phenolic compounds in acacia wood, and consequently in wines aged in contact with this wood species, had already been reported by other authors [42,45]. However, contrary to this fact other authors evidenced a pronounced enrichment of model wines in polyphenolic substances in those model wines treated with oak chips compared to those treated with acacia or cherry, due to the significantly higher amounts of total extractable polyphenols of oak wood [48].

It is worthy to note that the contact of wood by means of chips with white wines implied a decrease on the browning potential index, due to the release of phenolic compounds from wood to wines. Phenomena such as the precipitation of oxidized phenols, the formation of phenolic polymers and the antioxidant properties of some phenolic compounds, lead to increases in the stability of white wines to oxidation [11]. This fact should be also confirmed in case of the use of barrels where a gentle flow of oxygen is produced through the wood pores, since no references regarding this have been found. Among all types of wood tested, those wines aged with cherry chips exhibited the higher browning potential values which pointed it out as the most sensitive to oxidation [11]. This fact was in consonance to low content of oxidizable polyphenols that characterized cherry heartwood [42,49].

From a sensorial point of view, it seems that white wines aged with acacia chips showed major colour intensity compared to those wines treated with cherry or American and French oak wood. Indeed, comparing the values of colour difference (ΔE) between control wine and wines aged in contact with different wood chips species, only white wine aged with acacia wood chips showed values higher than two CIELAB units, which is detectable by human eyes [11], being CIELAB the color space defined by the International Commission on Illumination (CIE).

On the other hand, the use of cherry and acacia wood seemed to impart to white wines, less aggressiveness and less woody character than oak wood. Furthermore, contrary to oak wood, acacia barrels appeared to enhance the sweetness and honey tastes and pronounce the vanilla and spicy character of Malvazija wines. These facts resulted in better overall appreciation scores for white wines aged with acacia than for those aged with oak or other woods [11,50]. However, the typical aged flavours described as spicy or vanilla were not detected in Chardonnay wines, a more international grape variety, aged in both acacia barrels and acacia chips [10]. The reasons of these facts were the low content of volatile phenols such as eugenol and guaiacol, the low level of vanillin and the lack of other substances contributing to vanilla flavour such as *cis*- and *trans*-β-mehtyl-γ-octalactone detected in wines aged with acacia. On the other hand, the sweetness character of wines aged in acacia was also confirmed by other authors who found new aged flavours coming from the contact with acacia wood described as nutty, honeyed and toasted [10]. Based on the literature, the responsible compounds for these new aromas seem to be 2-acetyl pyrazine, 2-acetyl-3-methylpyrazine and 2-acetylthiazole, which were identified as distinctive of acacia wood due to their powerful sensorial features described as toasty, popcorn and nutty [51].

Despite the fact that it seems that there is a clear preference for acacia wood against oak or other woods to carry out the ageing of white wines, crucial aspects related to ageing treatment (barrels or chips) and ageing time have not been elucidated yet. Therefore, our research group compared the traditional ageing technique of a Chardonnay wine by means of acacia barrels with the alternative practice of use acacia chips obtained from manufacturing acacia barrels. Both assays were performed in duplicate and acacia wood underwent a light toasting process. Sensorial properties of aged wines were monitored along four timing points, monthly for barrels during 4 months and weekly for chips during 4 weeks. After sampling was completed, wine control without acacia contact and all aged wines were submitted to a descriptive sensorial analysis to assess the best maturation process conditions with acacia wood, based on the sensorial properties.

In general terms, the emergence of new sensory features described from acacia wood, nutty, honeyed and toasted, appeared as a consequence of the acacia ageing of wines. These new pleasant sensorial features were clearly accentuated in wines aged in barrels during three and four-month ageing processes [10]. Furthermore, the new aged attributes were in good balance with varietal aromas coming from Chardonnay wine varieties. The perception of acacia ageing notes was imperceptible for those wines with less contact with acacia, a one month of stay in barrels and one week macerated with chips, which preserved better the fresh, fruity and varietal characters of young wine. The better retention of fruity characteristics of the original wine by the use of chips rather than barrels has been previously reported by other authors [38,52].

Among wine samples treated with acacia chips, those that were macerated for the three weeks ageing process obtained higher scores by the panellists. Meanwhile, wines that were in contact with acacia during three and four months were the most regarded samples aged in barrels. Average values of each descriptor from the sensorial analysis of the highest scored aged wines in comparison with a control are shown by means of spider web diagrams in Figure 1. Compared to the use of barrels, the treatment with chips was scored lower by panellists. The use of chips led to a higher acidity and the attributes from acacia wood were timidly increased, leading to lower scores of taste intensity, taste quality and global quality. On the other hand, wines aged in barrels for three and four months exhibited a sensory profile different from the control. They resulted in high complexity due to the good balance between the varietal features of Chardonnay wine and the emergence of the new clearly perceptible sensory notes described for acacia wood, such as nutty, honeyed, and toasty, as a consequence of their contact with acacia barrels.

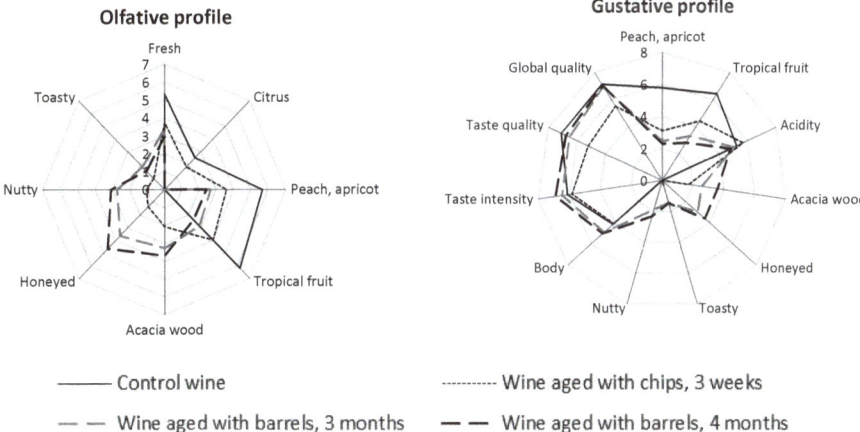

Figure 1. Olfative and gustative profiles of aged wines with the acacia treatments best scored in comparison with sensorial profiles of control wines.

With the aim of looking for other alternative woods to carry out the ageing processes that might create unique flavours, an exploratory survey evaluated 12 wood species from New Zealand in comparison with American oak wood [53]. These wood species had never before used as flavourants in wine: Matai, Feijoa, Macrocarpa, Pohutukawa, Radiata pine, Totara, Kahikatea, Rimu, Cherry beech, Silver beech and Manuka were light and dark toasted in the manner of oak barrels and then used to infuse Chardonnay wines. Each wood showed different behaviours in terms of sensorial properties. In general terms, typical oaked wine descriptors such as woody, smoky, vanilla and buttery were provided by these woods. However, other unattractive sensorial features described as earthy, sappy, resin, paint stripper or pencil sharpening's were detected in one of the wood treatments. Based on a hedonic consumer trial, it was concluded that each wood could generate a flavour liked by some consumers, so wood species as yet untested may be useful in conferring unique flavours from a particular geographical region.

4. Conclusions

Although the ageing stage has been exclusive to red wines, in recent years, the effects of wood contact on white wines have been addressed by several scientific researchers. New white winemaking techniques such as fermentation with oak chips and barrels, or the ageing on lees entail sensorial advantages due to the highest quantities of volatile oak-extractable compounds. On the other hand, new wood species alternative to traditional oak wood species offer the possibility to flavour white wines in a different manner in which oak wood does. Therefore, the treatment of white wines with wood by means of technological innovations or the use of alternative woods from oak, not only might drive to an added-value on quality but also to the acquisition of unique flavours in white wines leading to diversification of market opportunities. However, in the majority of cases, scientific studies have been done with minority grape varieties, so further studies should be conducted with more international varieties, to address the effects of wood contact with white wines more deeply.

Author Contributions: The three authors contributed to the writing and correction of the paper. The conceptualization, M.S.P-C.; writing-original draft preparation, M.E.A and M.C.D-M.; writing-review and editing, M.E.A.; Supervision, M.S.P-C and M.C.D-M. All authors red and approved the final manuscript.

Funding: The Spanish National Institute for Agricultural and Food Research and Technology (INIA) supported this work: Project RTA2014-00055-C03-02.

Acknowledgments: M.E.A. thanks University of Castilla-La Mancha for the postdoctoral contract (Access to the Spanish System of Science, Technology and Innovation (SECTI)).

Conflicts of Interest: Authors declare no conflict of interest.

References

1. Sánchez-Palomo, E.; Díaz-Maroto, M.C.; González Viñas, M.A.; Pérez-Coello, M.S. Aroma enhancement in wines from different grape varieties using exogenous glycosidases. *Food Chem.* **2005**, *92*, 627–635. [CrossRef]
2. Sánchez-Palomo, E.; Pérez-Coello, M.S.; Díaz-Maroto, M.C.; González Viñas, M.A.; Cabezudo, M.D. Contribution of free and glycosidically-bound volatile compounds to the aroma of muscat "a petit grains" wines and effect of skin contact. *Food Chem.* **2006**, *95*, 279–289. [CrossRef]
3. Sánchez-Palomo, E.; González Viñas, M.A.; Díaz-Maroto, M.C.; Soriano-Pérez, A.; Pérez-Coello, M.S. Aroma potential of Albillo wines and effect of skin-contact treatment. *Food Chem.* **2007**, *103*, 631–640. [CrossRef]
4. Jiménez Moreno, N.; Ancín Azpilicueta, C. Binding of oak volatile compounds by wine lees during simulation of wine ageing. *LWT-Food Sci. Technol.* **2007**, *40*, 619–624. [CrossRef]
5. Liberatone, M.T.; Pati, S.; Del Nobile, M.A.; La Notte, E. Aroma quality improvement of Chardonnay White wine by fermentation and ageing in barrique on lees. *Food Res. Int.* **2010**, *43*, 996–1002. [CrossRef]
6. Chatonnet, P.; Dubourdieu, D.; Boidron, J.N. Incidence des conditions de fermentation at d'elevage des vinsblanc secs en barriques sur leur composition in substances cèdèes par le bois de chêne. *Sci. Aliments* **1992**, *12*, 665–685.
7. Guchu, E.; Díaz-Maroto, M.C.; Pérez-Coello, M.S.; González Viñas, M.A.; Cabezudo, M.D. Volatile composition and sensory characteristics of Chardonnay wines treated with American and Hungarian oak chips. *Food Chem.* **2006**, *99*, 350–359. [CrossRef]

8. Baiano, A.; Varva, G.; De Gianni, A.; Viggiani, I.; Terracone, C.; Del Nobile, M.A. Influence of type of amphora on physico-chemical properties and antioxidant capacity of "Falanghina" white wines. *Food Chem.* **2014**, *146*, 226–233. [CrossRef] [PubMed]
9. Lukic, I.; Jedrejcic, N.; KovacevicGanic, K.; Staver, M.; Persuric, D. Phenolic and aroma composition of white wines produced by prolonged maceration and maturation in wooden barrels. *Food Tech. Biotech.* **2015**, *53*, 407–418.
10. Alañón, M.E.; Marchante, L.; Alarcón, M.; Díaz-Maroto, I.J.; Pérez-Coello, S.; Díaz-Maroto, M.C. Fingerprints of acacia aging treatments by barrels or chips based on volatile profile, sensorial properties, and multivariate analysis. *J. Sci. Food Agric.* **2018**, *98*, 5795–5806. [CrossRef] [PubMed]
11. Delia, L.; Jordão, A.M.; Ricardo Da Silva, J.M. Influence of different wood chips species (oak, acacia and cherry) used in a short period of aging on the quality of Encruzado white wines. *Mitt. Klosterneubg.* **2017**, *67*, 84–96.
12. Fernández de Simón, B.; Esteruelas, E.; Muñoz, A.M.; Cadahía, E.; Sanz, M. Volatile compounds in acacia, chestnut, cherry, ash, and oak woods, with a view to their use in cooperage. *J. Agric. Food Chem.* **2009**, *57*, 3217–3227. [CrossRef] [PubMed]
13. Alañón, M.E.; Castro-Vázquez, L.; Díaz-Maroto, M.C.; Pérez-Coello, M.S. Aromatic potential of *Castanea sativa* Mill. Compared to *Quercus* species to be used in cooperage. *Food Chem.* **2012**, *130*, 875–881. [CrossRef]
14. Alarcón, M.; Díaz-Maroto, M.C.; Pérez-Coello, M.S.; Alañón, M.E. Isolation of natural flavoring compounds from cooperage woods by pressurized hot water extraction (PHWE). *Holzforschung* **2018**. [CrossRef]
15. Guchu, E.; Díaz-Maroto, M.C.; Díaz-Maroto, I.J.; Vila-Lameiro, P.; Pérez-Coello, M.S. Influence of the species and geographical location on volatile composition of Spanish oak wood (*Quercus petraea* Liebl. and *Quercus robur* L.). *J. Agric. Food Chem.* **2006**, *54*, 3062–3066. [CrossRef] [PubMed]
16. Díaz-Maroto, M.C.; Guchu, E.; Castro-Vázquez, L.; de Torres, C.; Pérez-Coello, M.S. Aroma-active compounds of American, French, Hungarian and Russian oak Woods, studied by GC-MS and GC-O. *Flavour Fragr. J.* **2008**, *23*, 93–98. [CrossRef]
17. Prida, A.; Puech, J.L. Influence of geographical origin and botanical species on the content of extractives in American, French, and East European oak woods. *J. Agric. Food Chem.* **2006**, *54*, 8115–8126. [CrossRef] [PubMed]
18. Díaz-Maroto, I.J.; Vila-Lameiro, P.; Guchu, E.; Díaz-Maroto, M.C. A comparison of the autecology of *Quercus robur* L. and *Q. pyrenaica* Wild. present habitat in Galicia, NW Spain. *Forestry* **2007**, *80*, 223–239. [CrossRef]
19. Spillman, P.J.; Sefton, M.A.; Gawel, R. The contribution of volatile compounds derived during oak barrel maturation to the aroma of a Chardonnay and Cabernet Sauvignon wine. *Aust. J. Grape Wine Res.* **2004**, *10*, 227–235. [CrossRef]
20. Herjavec, S.; Jeromel, A.; Da Silva, A.; Orlic, S.; Redzepovic, S. The quality of white wines fermented in Croatian oak barrels. *Food Chem.* **2007**, *100*, 124–128. [CrossRef]
21. Prida, A.; Chatonnet, P. Impact of oak-derived compounds on the olfactory perception of barrel-aged wines. *Am. J. Enol. Vitic.* **2010**, *61*, 408–413.
22. Herrero, P.; Sáenz-Navajas, M.P.; Avizcuri, J.M.; Culleré, L.; Balda, P.; Antón, E.C.; Ferreira, V.; Escudero, A. Study of chardonnay and Sauvignon blanc wines form D.O.Ca Rioja (Spain) aged in different French oak wood barrels: Chemical and aroma quality aspects. *Food Res. Int.* **2016**, *89*, 227–236. [CrossRef] [PubMed]
23. Rodríguez-Nogales, J.M.; Fernández-Fernández, E.; Vila-Crespo, J. Characterisation and classification of Spanish Verdejo young white wines by volatile and sensory analysis with chemometric tools. *J. Sci. Food Agric.* **2009**, *89*, 1927–1935. [CrossRef]
24. Gutiérrez Afonso, V.L. Sensory descriptive analysis between white wines fermented with oak chips and in barrels. *J. Food Sci.* **2002**, *67*, 2415–2419. [CrossRef]
25. Aleixandre, J.L.; Padilla, A.I.; Navarro, L.L.; Suria, A.; García, M.; Álvarez, I. Optimisation of making barrel-fermented dry Muscatel wines. *J. Agric. Food Chem.* **2003**, *51*, 1889–1893. [CrossRef] [PubMed]
26. Nunes, P.; Muxagata, S.; Correia, A.C.; Nunes, F.M.; Cosme, F.; Jordao, A.M. Effect of oak wood barrel capacity and utilization time on phenolic and sensorial profile evolution of an Encruzado white wine. *J. Sci. Food Agric.* **2017**, *97*, 4847–4856. [CrossRef] [PubMed]
27. Ortega-Heras, M.; González-Sanjosé, M.L.; González-Huerta, C. Consideration of the influence of ageing process, type of wine and oenological classic parameters on the levels of wood volatile compounds present in red wines. *Food Chem.* **2007**, *103*, 1434–1448. [CrossRef]

28. Karbowiak, T.; Gougeon, R.D.; Alinc, J.B.; Brachais, L.; Debeaufort, F.; Voilley, A.; Chassagne, D. Wine oxidation and the role of cork. *Crit. Rev. Food Sci. Nutr.* **2009**, *50*, 20–52. [CrossRef]

29. Pérez-Coello, M.S.; González-Viñas, M.A.; García-Romero, E.; Cabezudo, M.D.; Sanz, J. Chemical and sensory changes in white wines fermented in the presence of oak chips. *Int. J. Food Sci. Technol.* **2000**, *35*, 23–32. [CrossRef]

30. Sánchez-Palomo, E.; Alonso-Villegas, R.; Delgado, J.A.; González-Viñas, M.S. Improvement of Verdejo white wines by contact with oak chips at different winemaking stages. *LWT-Food Sci. Technol.* **2017**, *79*, 111–118. [CrossRef]

31. Zea, L.; Moyano, L.; Moreno, J.A.; Medina, M. Aroma series as fingerprints for biological ageing in fino sherry-type wines. *J. Sci. Food Agric.* **2007**, *87*, 2319–2326. [CrossRef]

32. Cadwallader, K.R. Potent odorants in hickory and mesquite smokes and liquid smoke extracts. In Proceedings of the Annual Meeting of the Institute of Food Technologies, New Orleans, LA, USA, 22–26 June 1996; pp. 34–36.

33. Sunao, M.; Ito, T.; Hiroshima, K.; Sato, M.; Uehara, T.; Ohno, T.; Watanabe, S.; Takahashi, H.; Hashizume, K. Analysis of volatile phenolic compounds responsible for 4-vinylguaiacol-like odor characteristics of sake. *Food Sci. Technol. Res.* **2016**, *22*, 111–116. [CrossRef]

34. Escudero, A.; Campo, E.; Fariña, L.; Cacho, J.; Ferreira, V. Analytical characterization of the aroma of five Premium red wines. Insights into the role of odor families and the concept of fruitiness of wines. *J. Agric. Food Chem.* **2007**, *55*, 4501–4510. [CrossRef] [PubMed]

35. López, R.; Aznar, M.; Cacho, J.; Ferreira, V. Determination of minor and trace volatile compounds in wine by solid-phase extraction and gas chromatography with mass spectrometric detection. *J. Chromatogr. A* **2002**, *966*, 167–177. [CrossRef]

36. Etievant, P.X. Wine. In *Volatile Compounds in Foods and Beverages*; Marse, H., Ed.; Marcek Dekker, Inc.: New York, NY, USA, 1991; pp. 483–587.

37. Cutzach, I.; Chatonnet, P.; Dubourdieu, D. Study of the formation mechanisms of some volatile compounds during aging of sweet fortified wines. *J. Agric. Food Chem.* **1999**, *47*, 2837–2846. [CrossRef] [PubMed]

38. Alañón, M.E.; Schumacher, R.; Castro-Vázquez, L.; Díaz-Maroto, M.C.; Pérez-Coello, M.S. Enological potential of chestnut wood for aging Tempranillo wines Part I: Volatile compounds and sensorial properties. *Food Res. Int.* **2013**, *51*, 325–334. [CrossRef]

39. Alañón, M.E.; Schumacher, R.; Castro-Vázquez, L.; Díaz-Maroto, M.C.; Hermosín, I.; Pérez-Coello, M.S. Enological potential of chestnut wood for aging Tempranillo wines Part II: Phenolic compounds and chromatic characteristics. *Food Res. Int.* **2013**, *51*, 536–546. [CrossRef]

40. Chinnici, F.; Natali, N.; Bellachioma, A.; Versari, A.; Riponi, C. Changes in phenolic composition of red wines aged in cherry wood. *LWT-Food Sci. Technol.* **2015**, *60*, 977–984. [CrossRef]

41. Chinnici, F.; Natali, N.; Sonni, F.; Bellachioma, A.; Riponi, C. Comparative changes in color features and pigment composition of red wines aged in oak and cherry wood casks. *J. Agric. Food Chem.* **2011**, *59*, 6575–6582. [CrossRef] [PubMed]

42. De Rosso, M.; Paniguel, A.; Dalla Vedova, A.; Stella, L.; Flamini, R. Changes in chemical composition of a red wine aged in acacia, cherry, chestnut, mulberry and oak wood barrels. *J. Agric. Food Chem.* **2009**, *57*, 1915–1920. [CrossRef] [PubMed]

43. Fernández de Simón, B.; Martínez, J.; Sanz, M.; Cadahía, E.; Esteruelas, E.; Muñoz, A.M. Volatile compounds and sensorial characterization of red wine aged in cherry, chestnut, false acacia, ash and oak wood barrels. *Food Chem.* **2014**, *147*, 346–356. [CrossRef] [PubMed]

44. Fernández de Simón, B.; Sanz, M.; Cadahía, E.; Martínez, J.; Muñoz, A.M. Polyphenolic compounds as chemical markers of wine ageing in contact with cherry, chestnut, false acacia, ash and oak wood. *Food Chem.* **2014**, *143*, 66–76. [CrossRef] [PubMed]

45. Sanz, M.; Fernández de Simón, B.; Esteruelas, E.; Muñóz, A.M.; Cadahía, E.; Hernández, T.; Estrella, I.; Martínez, J. Polyphenols in red wine aged in acacia (*Robinia pseudoacacia*) and oak (*Quercus petraea*) wood barrels. *Anal. Chim. Acta* **2012**, *732*, 83–90. [CrossRef] [PubMed]

46. Gortzi, O.; Metaxa, X.; Mantanis, G.; Lalas, S. Effect of artificial ageing using different wood chips on the antioxidant activity, resveratrol and catechin concentration, sensory properties and color of two Greek red wines. *Food Chem.* **2013**, *141*, 2887–2895. [CrossRef] [PubMed]

47. Kyraleou, M.; Kallithraka, S.; Chira, K.; Tzanakouli, E.; Ligas, I.; Kotseridis, Y. Differentiation of wines treated with wood chips based on their phenolic content, volatile composition, and sensory parameters. *J. Food Sci.* **2015**, *80*, C2701–C2710. [CrossRef] [PubMed]

48. Psarra, C.; Gortzi, O.; Dimitirs, P.M. Kinetics of polyphenol extraction from wood chips in wine models solutions: Effect of chip amount and botanical species. *J. Inst. Brew.* **2015**, *121*, 207–212. [CrossRef]

49. Alañón, M.E.; Castro-Vázquez, L.; Díaz-Maroto, M.C.; Hermosín-Gutiérrez, I.; Gordon, M.H.; Pérez-Coello, M.S. Antioxidant capacity and phenolic composition of different woods used in cooperage. *Food Chem.* **2011**, *129*, 1584–1590. [CrossRef]

50. Kozlovic, G.; Jeromel, A.; Maslov, L.; Pollnitz, A.; Orlic, S. Use of acacia barrique barrels—Influence on the quality of Malvazija from Istria wines. *Food Chem.* **2010**, *120*, 698–702. [CrossRef]

51. Culleré, L.; Fernández de Simón, B.; Cadahía, E.; Ferreira, V.; Hernández Orte, P.; Cacho, J. Characterization by gas chromatography-olfactometry of the most odor-active compounds in extracts prepared from acacia, chestnut, cherry, ash and oak woods. *LWT-Food Sci. Technol.* **2013**, *53*, 240–248. [CrossRef]

52. Pérez-Coello, M.S.; Sánchez, M.A.; García, E.; González-Viñas, M.S.; Sanz, J.; Cabezudo, M.D. Fermentation of white wines in the presence of wood chips of American and French oak. *J. Agric. Food Chem.* **2000**, *48*, 885–889. [CrossRef] [PubMed]

53. Young, O.A.; Kaushal, M.; Robertson, J.D.; Burns, H.; Nunns, S.J. Use of species other than oak to flavor wine: An exploratory survey. *J. Food Sci.* **2010**, *75*, S490–S498. [CrossRef] [PubMed]

Article

Effects of Fining Agents, Reverse Osmosis and Wine Age on Brown Marmorated Stink Bug (*Halyomorpha halys*) Taint in Wine

Pallavi Mohekar, James Osborne and Elizabeth Tomasino *

Department of Food Science & Technology, Oregon State University, Corvallis, OR 97331, USA;
pallavi.mohekar@gmail.com (P.M.); james.osborne@oregonstate.edu (J.O.)
* Correspondence: elizabeth.tomasino@oregonstate.edu; Tel.: +1-541-737-4866

Received: 3 January 2018; Accepted: 25 January 2018; Published: 1 March 2018

Abstract: *Trans*-2-decenal and tridecane are compounds found in wine made from brown marmorated stink bug (BMSB)-contaminated grapes. The effectiveness of post-fermentation processes on reducing their concentration in finished wine and their longevity during wine aging was evaluated. Red wines containing *trans*-2-decenal were treated with fining agents and put through reverse osmosis filtration. The efficacy of these treatments was determined using chemical analysis (MDGC-MS) and sensory descriptive analysis. Tridecane and *trans*-2-decenal concentrations in red and white wine were determined at bottle aging durations of 0, 6, 12 and 24 months using MDGC-MS. Reverse osmosis was found to be partially successful in removing *trans*-2-decenal concentration from finished wine. While tridecane and *trans*-2-decenal concentrations decreased during bottle aging, post-fermentative fining treatments were not effective at removing these compounds. Although French oak did not alter the concentration of tridecane and *trans*-2-decenal in red wine, it did mask the expression of BMSB-related sensory characters. Because of the ineffectiveness of removing BMSB taint post-fermentation, BMSB densities in the grape clusters should be minimized so that the taint does not occur in the wine.

Keywords: *trans*-2-decenal; tridecane; MDGC-MS; red wine; Pinot noir

---

1. Introduction

Brown marmorated stink bug (BMSB) contamination in grape clusters can have a negative effect on wine quality [1,2]. BMSB is an invasive pest that is believed to have arrived into the United States from East Asia and is currently detected in 43 states. Globally, it is also found in Canada, Italy, Hungary and other European countries where wine has economic importance [3,4]. When present in the vineyard, the pest can lower crop yield and effect quality. When present in grape cluster, it may enter wine processing where it can harm wine quality through the release of "BMSB taint" compounds. The chance of BMSB entering wine processing is increasing, as greater densities of BMSB are being observed in the vineyard [3,5,6]. In order to maintain wine quality, techniques are needed to minimize BMSB taint concentration in finished wine.

BMSB primarily secretes tridecane and *trans*-2-decenal when stressed [7–9]. Tridecane is an odorless compound and its effect on wine quality is currently unknown. *Trans*-2-decenal is considered to be the main component of BMSB taint due to its strong "green", "cilantro"-like aroma [2,10]. It has been shown to have a negative effect on red wine quality, significantly decreasing consumer preference at a concentration as low as 4.8 µg/L, the determined consumer rejection threshold (CRT) [1]. Above this concentration *trans*-2-decenal can add green, musty, herbal characteristics to wine which are not desirable [11]. Additionally, a reduction in favorable attributes such as dark fruit, red fruit and floral characteristics has also been observed [11]. Due to this negative impact of BMSB taint on wine quality and consumer preference, efforts are needed to minimize the concentration of these taint compounds in finished wine.

It has been shown that as low as three BMSB per cluster in the vineyard can result in finished wine with 2.02 μg/L of *trans*-2-decenal [12]. The same bug density may also result in *trans*-2-decenal concentration at or above the CRT when winemaking causes stress to BMSB, resulting in higher secretion of taint compounds. Additionally, previous work suggests that the presence of dead BMSB can also result in wine containing tridecane but not *trans*-2-decenal [12].

Modifications in wine making protocol may reduce BMSB taint concentrations in final wine, as winemaking processes are known to alter aroma composition. Alterations in harvesting, pressing and fermentation have shown potential in reducing BMSB taint in finished wine [12]. However, modification of wine processing may not be always appropriate as it can restrict the style of wine made from BMSB contaminated grapes. Additionally, process modification may not be sufficient against high BMSB densities and finished wine may still contain *trans*-2-decenal or tridecane. Therefore, post-fermentative measures are required to be able to produce a desired wine style while minimizing taint levels.

The wine industry relies on fining agents to correct wine faults and to improve wine quality [13–15]. Wine sensory characteristics such as flavor, color and mouthfeel can be adjusted by fining agents [14,16,17]. Fining agents are chosen for their affinity to unwanted compounds in wine through mechanisms such as hydrophobic interaction, hydrogen bonds, Vander Waals interaction and electrostatic interactions [13,18]. The fining agent, along with unwanted or taint compounds, are then removed by racking, centrifugation or filtering. Commonly used fining agents such as bentonite, gelatin, casein and activated charcoal have previously been used on taint compounds from lady bug and smoke exposure [13,15]. In these studies, oak was able to mask green aroma characteristics of lady beetle taint in red and white wine whereas activated charcoal and synthetic mineral were effective against smoke taint compounds.

In addition to fining agents, reverse osmosis filtration has also been explored as a viable option for taint removal [19]. In this process, selective taint removal can be achieved by carrying out filtration under pressure. Reverse osmosis is often combined with an adsorption or ion-exchange column to remove taint compounds more efficiently. This technique has been found to be successful in removing 4-ethylguaiacol and 4-ethylphenol from *Brettanomyces*-affected wine [20] using Amberlite XAD-16 HP resin and smoke taint compounds (guaiacol, 4-methylguaiacol, 4-ethylguaiacol and 4-ethylphenol) using a polystyrene-based adsorbent resin [19].

Another possible technique to reduce BMSB taint in wine is through aging. During aging, complex reactions occur that are known to change wine composition and sensory characteristics [21,22]. Since most wines are aged for at least a year, it is important to understand how BMSB taint is modified during aging. This information is important in order to assess the quality of an aged wine. Currently, there is no technique known to be effective against BMSB taint in finished wine and reduction techniques are needed to help deal with this spoilage issue when control of BMSB in the vineyard is ineffective.

2. Materials and Methods

Wines—Three different Pinot noir wines (PN1, PN2 and PN3) were used in this study. All wines were produced on a small scale at the Oregon State University research winery (Corvallis, OR, USA). PN1 and PN2 were produced from the same grapes sourced from a vineyard located in Oregon and under the same winemaking protocol. PN1 was made from grapes that did not contain any BMSB. PN2 on the other hand, was made from grapes to which BMSB were added prior to destemming at a density of three per cluster. PN2 therefore contained both taint compounds, tridecane and *trans*-2-decenal whereas PN1 was taint free. Both wines, PN1 and PN2 were made using a winemaking protocol as described in [12]. These wines were used to study the effect of aging and reverse osmosis filtration. PN3 was also produced without any addition of BMSB but the grapes were sourced from a different vineyard in Oregon. Winemaking procedure used to make PN3 was similar to the protocol given in [17]. *Trans*-2-decenal was added to this wine to study the effect of fining treatment, as not enough BMSB were available that year to make naturally tainted wines at the concentrations needed.

Fining agents—Fining agents and their dose levels were selected based on preliminary studies and/or manufacturer recommendations. In the end, five fining agents: gelatin (BBL, Div Becton Dickinson & Co., Sparks, MD, USA), egg albumin, potassium caseinate (Laffort USA, Petaluma, CA, USA), bentonite, yeast lees (Oenolees®, Laffort USA, Petaluma, CA, USA) and French oak bean, medium plus toast (StaVin Inc., Sausalito, CA, USA) were tested. Egg albumin solution was prepared in 1% NaCl using eggs from the local grocery store. Other fining agents were prepared in hot or cold water per manufacturer's instruction. The following dose levels were used for each fining agent: gelatin at 30 mg/L, egg albumin at 67 mg/L, potassium caseinate at 150 mg/L, bentonite at 75 mg/L, yeast lees at 150 mg/L and French Oak bean at 1.5 g/L The addition rate for French oak was based on the manufacturer's instruction for 50% new oak.

To run fining trials, Pinot noir with *trans*-2-decenal concentration of 30 µg/L was prepared. This was done by adding *trans*-2-decenal standard (50 mg/L made in 14% ethanol) into the base wine, PN3. The concentration of 30 µg/L was selected for fining treatment because it is significantly above *trans*-2-decenal CRT (4.8 µg/L) and has been shown to add green sensory characteristics associated with *trans*-2-decenal. Additionally, a significant proportion of consumers (78%) were seen to reject Pinot noir containing 30µg/L of *trans*-2-decenal [1].

Twenty-four hours after *trans*-2-decenal addition into PN3, fining agents were added. PN3 containing *trans*-2-decenal and fining agent were then stored at 4 °C for three days. At the end of three days, fining agents were removed by racking, wines were rebottled and stored for analysis at 4 °C. Twenty hours after racking, wines were analyzed using sensory descriptive analysis. At the same time, 40 mL sample of these racked wines was collected for *trans*-2-decenal quantification using MDGC-MS. Samples were stored in amber vials with PTFE lined caps (Sigma Aldrich, Darmstadt, Germany) at −18 °C until their analysis. All fining trials were conducted in triplicate.

Reverse osmosis—Two wines, PN1 and PN2 went through reverse osmosis filtration conducted by WineSecrets Corp. (Sebastopol, CA, USA). PN1 was treated by reverse osmosis as well as the BMSB tainted wine to determine the effect of reverse osmosis on aroma compounds not typically associated with BMSB. Therefore, both wines were treated in exactly the same manner. Reverse osmosis was performed on a lab scale Memstar unit with a CBC-5 carbon block filter cartridge (Pentair Pentek, Milwaukee, WI, USA) attachment. Feeding pump pressure was maintained between 1700–1800 kPa and total sample flow rate at 50 mL per minute. All wines that went through reverse osmosis were done in triplicate.

Aging—PN1 and PN2 wines were aged in 750 mL screw-cap closed (Stelvin, Amcor, CA, USA) bottles in a dark wine cellar located at Oregon State University at 13 °C for 1 year. At 0, 6 and 12 months, bottles were removed from the cellar and 40 mL samples were stored in amber vials with PTFE lined caps (Sigma Aldrich) at −18 °C for later analysis. All wines were aged in triplicate.

Chemical Analysis—Taint compounds were measured using a previously developed HS-SPME-MDGC-MS method [12].

Sensory Analysis—Descriptive analysis was used to determine the effect of fining treatment and reverse osmosis on wine sensory characteristics. Sixteen wine professionals (12 M, 4 F) from the Oregon wine industry participated in this study. Each panelist had more than 10 years of experience tasting wines. Consent was obtained from all panelists and the study was approved by Oregon State University's Internal Review Board (IRB). Descriptive analysis data was collected over three tasting sessions, each lasting two hours. Each of the three sessions was conducted in the morning. The third session was conducted in a different room but under similar light and temperature (21 ± 2 °C) conditions. Wines were served in INAO black glasses (International Organization for Standardization 1977) to remove any influence of color [23]. All samples were coded with a three digit random numbers and served in a random order.

At the start of each session, panelists were given a set of three wine samples; control, Pinot noir with *trans*-2-decenal at its CRT (4.8 µg/L), and Pinot noir with *trans*-2-decenal above its CRT (30 µg/L). This was done to familiarize panelists with the taint compound and its effect on Pinot noir aroma and flavor. These wines were prepared an hour before the tasting session by adding a *trans*-2-decenal standard (50 µg/L prepared in 14% ethanol). Wines were served in three sets containing five samples each. To avoid the effect of fatigue, panelists were given a one-minute break after each wine and five minutes after each set. Panelists were requested to rinse their palate and eat a cracker during each break to minimize any carryover effect.

Samples were evaluated for ten aromas (dark fruit, earthy, herbal, musty, red fruit, floral, fresh green, spice) and three flavors (fruit density, green, and spice). These attributes have been used previously to evaluate the effect of *trans*-2-decenal on Pinot noir quality [11]. Each attribute was rated on a 100 mm visual analog scale with indented word anchors, none and extreme. Panelists were allowed to rate any other attribute they thought was relevant to describe these samples, to avoid any dumping effect [24].

Statistical Analysis—Any differences in *trans*-2-decenal concentration between PN3, with and without fining agents was analyzed using one-way ANOVA and Dunnett's test. Descriptive analysis data was analyzed using mixed model ANOVA to determine consensus among the assessors for each attribute and wine. The fixed effect was wine and the random effects were panelists and replication. Canonical variate analysis (CVA) was used to explore the separation between wine treatments [25,26]. Significant differences during aging were analyzed using one way Analysis of Variance (ANOVA) and Tukey's HSD. All analyses were conducted using XLSTAT-Pro 2015 (Addinsoft, New York, NY, USA).

3. Results

3.1. Chemical Analysis

Trans-2-decenal was added to PN3 and was present at a concentration of 23.88 µg/L prior to the addition of any fining agents. After fining, no significant differences were found between *trans*-2-decenal concentrations in base wine and fined or oaked wines ($p < 0.05$, Dunnett's). Treatment of PN2 wines by reverse osmosis reduced the *trans*-2-decenal concentration in wine from 2.02 µg/L to 1.82 µg/L (*t*-test, *p*-value < 0.05), a 10% reduction of the compound. Chemical analysis of the aged PN2 wines showed a decrease in tridecane concentration in PN2 wine during bottle aging (Figure 1). *Trans*-2-decenal was 2.02 µg/L at 0 months and decreased to undetectable by 6 months.

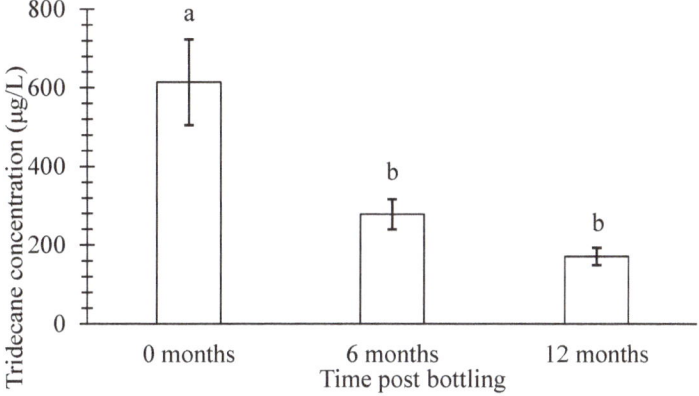

Figure 1. Tridecane concentration in Pinot noir made from BMSB containing grapes, PN2 at 0, 6 and 12 months of bottle aging (*n* = 3). Means with the same letter are not significantly different from each other (Tukey's HSD, α: 0.05).

3.2. Sensory Analysis

The effect of fining agent on reducing the sensory impact of *trans*-2-decenal was evaluated using eight wines. Six wines relating to the different fining treatments were evaluated. The other two wines were PN3, with and without *trans*-2-decenal. A significant interaction between panel and wine treatment was observed for spice flavor. This indicates inconsistency in the use of intensity scale or interpretation of this attribute by wine professional [24]. The effect of such an interaction can result from inherent anatomical differences in panelists. There is also the possibility of the term "spice" being too generic and therefore being interpreted differently by each panelist. Given that the interaction exists, spice flavor was excluded from CVA analysis. ANOVA of the sensory descriptors showed a significant difference in red fruit aroma between wine treatments (F value = 3.07, df = 7, *p*-value = 0.03). However, no difference in *trans*-2-decenal-related attributes, such as green, musty and earthy, were found. The only observed difference (a change in red fruit aroma) is typically considered as a side effect of using fining agents [27].

CVA analysis (Figure 2) showed a clear separation between the wines. Three distinct groups were visible:

Group 1—PN3 without *trans*-2-decenal
Group 2—French oak treatment
Group 3—remaining six wines (PN3 with *trans*-2-decenal and the other five fining treatments).

PN3 without *trans*-2-decenal was mainly characterized by fruity aroma and flavor. Wine with French oak treatment showed a strong spice aroma but no *trans*-2-decenal-related attributes. Since the chemical analysis of oak treated wines did not show any change in *trans*-2-decenal concentration the effect of oak is most likely a masking effect.

The two wines that underwent reverse osmosis (RO) were assessed by sensory analysis. PN3 with and without *trans*-2-decenal were also included in the analysis for comparison. Thus, in the second CVA analysis, a total of four wines were analyzed: (1) PN1 (made from BMSB free grapes) after RO; (2) PN2 (wine made from BMSB added grapes) after RO; (3) PN3 with *trans*-2-decenal and (4) PN3 without *trans*-2-decenal. The last two wines were included in the analysis in order to provide a means to compare the effect of reverse osmosis filtration and high levels of *trans*-2-decenal on wine aromatics.

A significant interaction was observed between panel and wine for green flavor. Using the same justification as before, green flavor was removed from CVA analysis. The results of CVA analysis are shown in Figure 3. Three groups were seen on a CVA plot:

Group 1—PN3 with *trans*-2-decenal
Group 2—PN3 without *trans*-2-decenal
Group 3—PN1 and PN2 after RO.

These three groups show clear differentiation between wine without *trans*-2-decenal, with *trans*-2-decenal and those wines that went through RO.

As with the previous sensory analysis, PN3 without *trans*-2-decenal was perceived as fruity and floral. The addition of *trans*-2-decenal brought out green, musty herbal characteristics in the same wine. None of these negative characteristics were found in PN1 and PN2 treated with RO. After reverse osmosis, PN2 was mainly characterized by earthy notes but none of the more pronounced attributes associated with *trans*-2-decenal such as green, musty and herbal were observed.

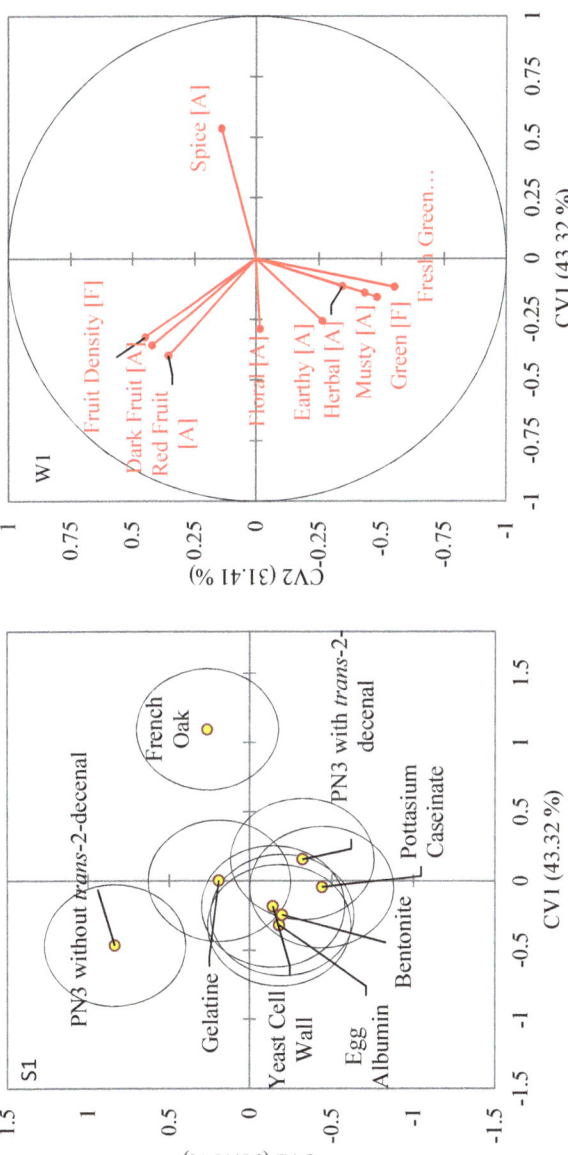

Figure 2. Separation between PN3 without *trans*-2-decenal, PN3 with *trans*-2-decenal (23.88 μg/L) and PN3 containing *trans*-2-decenal (23.88 μg/L) when treated with fining agents (Gelatin, Bentonite, Yeast cell wall, Potassium caseinate, Egg albumin, French oak). Wines are positioned using the centroids. Circles represent 95% confidence intervals surrounding the wine means. Vectors for sensory terms (A = aroma, F = in mouth flavor) are in W1 and scores for wines are in S1. Significant differences for wines are for circles that do not touch in S1.

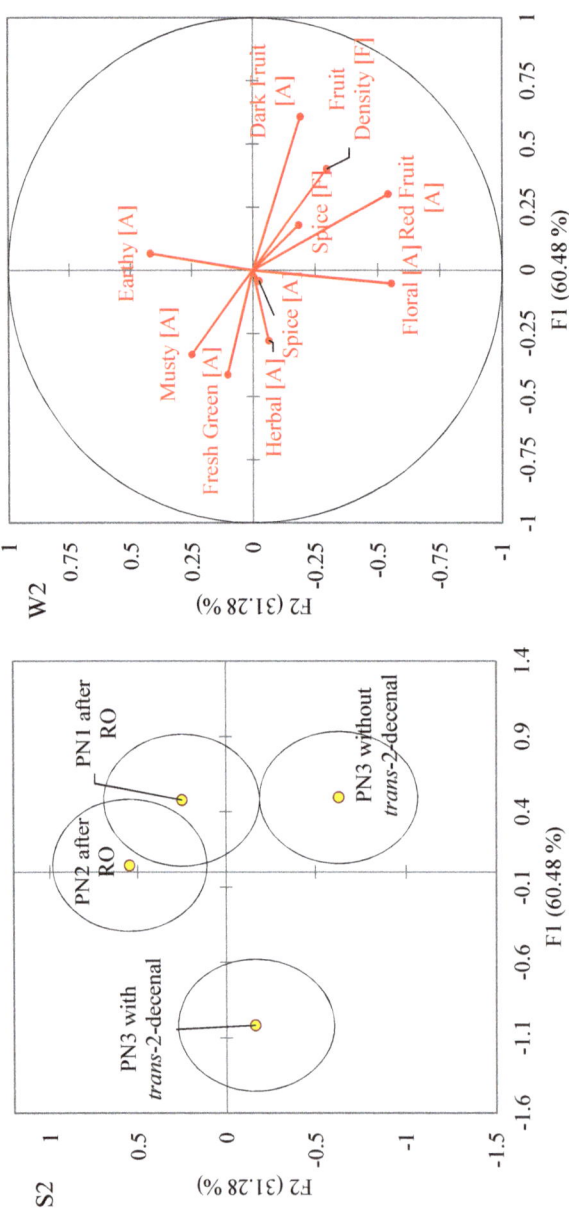

Figure 3. Separation among wines that went through reverse osmosis (RO) filtration: PN1 (wine made from *BMSB free* grapes), PN2 (wine made from BMSB containing grapes and contains 1.82 µg/L of *trans*-2-decenal) and wines that did not go through reverse osmosis (RO) filtration: PN3 without added *trans*-2-decenal, PN3 with *trans*-2-decenal (23.88 µg/L) by CVA. Wines are positioned using the centroids. Circles represent 95% confidence intervals surrounding the wine means. Vectors for sensory terms (A = aroma, F = in mouth flavor) are in W2 and scores for wines are in S2. Significant differences for wines are for circles that do not touch in S2.

4. Discussion

BMSB taint in wine can be detrimental to wine quality but little is known regarding ways to remove this taint from wine. While fining agents are often used during winemaking to remove wine taints, the fining agents evaluated in the present study were incapable of removing *trans*-2-decenal in BMSB tainted wines. Additional work is needed to determine the underlying factors for this result. One potential reason may be that the fining agents have only weak binding ability with *trans*-2-decenal. Alternatively, the fining agents may have higher affinity for phenolic compounds or other aroma compounds compared to *trans*-2-decenal. This explanation seems likely given that most fining agents are known to bind with non-volatile components in wine such as proteins and phenolic compounds [14,16,18,28]. Their interaction with volatile compounds is considered to be a secondary binding action and an undesirable effect that can be exploited for taint reduction [13,14,28,29].

The addition of French oak chips was the only treatment to have an impact on BMSB taint in the wines with a masking effect being observed during sensory analysis. Previous work on lady bug taint has also reported a similar masking effect of oak addition [15]. Wines treated with fining agents were all described with similar characteristics as the *trans*-2-decenal wines. Overall, the results of sensory evaluation agree with the conclusion of the chemical analysis data, namely, that fining agents failed to remove *trans*-2-decenal from wine.

Reverse osmosis resulted in a slight reduction of *trans*-2-decenal of 0.2 μg/L. This is minimal and would likely only be effective in changing sensory perception if the final concentration after fining was near *trans*-2-decenal CRT. However, this result indicates the ability of reverse osmosis to reduce *trans*-2-decenal, which was not found with other treatments. Therefore, with additional improvements reverse osmosis may prove to be a viable option for BMSB taint management. The use of other semipermeable membranes, adsorption/ion exchange column, pressure and flow rate should be investigated to remove greater amounts of *trans*-2-decenal.

The impact of RO on the sensory characteristics of *trans*-2-decenal are unclear. BMSB associated aromas including green, musty and herbal where not found in RO wines, but an earthy aroma was described in these wines. Previous sensory analysis conducted on wines with 5 μg/L *trans*-2-decenal also were found to be "earthy" [11]. However, wine after RO was described as different from the control (PN3 without *trans*-2-decenal) so it is unclear if this would be problematic to wine quality. The usage of RO to reduce BMSB warrants further research.

Aging appeared to be the most effective treatment to reduce *trans*-2-decenal and tridecane in BMSB tainted wines. A longer aging period may be preferable in wines containing BMSB taint since the aging process appears to naturally decrease their levels. The decrease in BMSB taint post bottling is likely to be a result of aging-related reactions occurring in wine such as hydrolysis, component degradation, condensation and reduction reactions [21,22,30]. These reactions can modify existing compounds or generate new ones.

Aldehydes, such as *trans*-2-decenal, are highly reactive and can bind with a number of different compounds such as SO_2 or phenolic compounds [31,32]. Prior work has shown that reductive conditions during bottle aging can cause aldehydes to decrease in wines as they change to their corresponding alcohol [14,33]. 1-Decanol, the corresponding alcohol for *trans*-2-decanal, has a detection threshold (5 mg/L) that is much greater than *trans*-2-decenal [34]. Therefore, as the wine ages and *trans*-2-decenal decreases the effects of 1-decanol may be minimal. However, if wines are matured under oxidative condition, acetals can form as result of reaction between aldehydes and alcohol [14,35]. Additional research is needed to better understand the transformation of *trans*-2-decenal during aging. This taint may only be problematic for young wines, as *trans*-2-decenal was undetectable after 6 months. However it is unknown if the transformation products of *trans*-2-decenal may also be problematic to wine quality.

The corresponding sensory impact of BMSB taint during aging also needs to be evaluated. This will estimate the effect after compounds released by BMSB have undergone aging-related changes. We did not have enough wine to conduct sensory tests on the aged wines but this is needed as preliminary sensory tests showed that BMSB taint wines were reduced in 1 year aged wines, but wines were very different from the control wines.

5. Conclusions

Corrective measures are of significant importance for wine containing BMSB taint. They provide the last option for winemakers to correct wine faults. Taking this into consideration, this study evaluated the effectiveness of reverse osmosis, commonly used fining agents, and bottle aging on BMSB taint compound. While fining agents proved ineffective at removing BMSB taint, reverse osmosis showed promise as *trans*-2-decenal concentrations in wine were reduced by 10% and resulted in improvements in the wines sensory characteristics. In addition to reverse osmosis, oak addition can also be considered while dealing with BMSB taint as it successfully masked the green characteristics associated with *trans*-2-decenal in wine. Finally, aging wines for extended periods of time may offer the best strategy for reducing BMSB taint in red wines and is also a process that most wines will undergo as part of the winemaking process. Taken together, wine containing BMSB can be treated by one of three options, addition of French oak, reverse osmosis or aging. Overall, the outcome of this study provides means of dealing with BMSB taint in finished wine. This information is important to minimize the impact of BMSB taint and maintain wine quality.

Acknowledgments: The authors would like to thank Oregon State University, Oregon Wine Research Institute and USDA-NIFA-SCRI #2011-51181-30937 and USDA # 59-5358-4-016 for financial support. We would also like to thank Stanvin for providing oak samples and WineSecrets for running reverse osmosis. Finally, we would like to thank the wine consumers and Oregon wine professionals for their participation in sensory evaluation.

Author Contributions: Elizabeth Tomasino and Pallavi Mohekar conceived and designed the experiments; Pallavi Mohekar performed the experiments and analyzed the data; Pallavi Mohekar and Elizabeth Tomasino wrote the paper, with James Osborne providing editing support.

Conflicts of Interest: "The authors declare no conflict of interest."

References

1. Mohekar, P.; Lim, J.; Lapis, T.; Tomasino, E. Consumer rejection thresholds of trans-2-decenal in Pinot noir: Linking wine quality to threshold segmentation. In Proceedings of the Institute of Food Technology Annual Meeting, New Orleans, LA, USA, 22–24 June 2014.
2. Tomasino, E. Impact of Brown Marmorated Stinkbug on Pinot noir Wine Quality. In Proceedings of the 64th ASEV National Conference, Monterey, CA, USA, 24–28 June 2013.
3. Haye, T.; Gariepy, T.; Hoelmer, K.; Rossi, J.-P.; Streito, J.-C.; Tassus, X.; Desneux, N. Range expansion of the invasive brown marmorated stinkbug, *Halyomorpha halys*: An increasing threat to field, fruit and vegetable crops worldwide. *J. Pest Sci.* **2015**, 1–9. [CrossRef]
4. Lee, D.-H. Current status of research progress on the biology and management of *Halyomorpha halys* (Hemiptera: Pentatomidae) as an invasive species. *Appl. Entomol. Zool.* **2015**, *50*, 277–290. [CrossRef]
5. Basnet, S. Biology and Pest Status of Brown Marmorated Stink Bug (Hemiptera: Pentatomidae) in Virginia Vineyards and Raspberry Plantings. Ph.D. Thesis, Virginia Polytechnic Institute and State University, Blacksburg, VA, USA, 2014.
6. Smith, J.R.; Hesler, S.P.; Loeb, G.M. Potential Impact of *Halyomorpha halys* (Hemiptera: Pentatomidae) on Grape Production in the Finger Lakes Region of New York. *J. Entomol. Sci.* **2014**, *49*, 290–303. [CrossRef]
7. Baldwin, R.L.; Zhang, A.; Fultz, S.W.; Abubeker, S.; Harris, C.; Connor, E.E.; Van Hekken, D.L. Hot topic: Brown marmorated stink bug odor compounds do not transfer into milk by feeding bug-contaminated corn silage to lactating dairy cattle. *J. Dairy Sci.* **2014**, *97*, 1877–1884. [CrossRef] [PubMed]
8. Mohekar, P.; Tomasino, E.; Wiman, N.G. Defining defensive secretions of brown marmorated stink bug, *Halyomorpha halys*. In Proceedings of the Entomology Society of American Annual Meeting, Minneapolis, MN, USA, 15–18 November 2015.
9. Solomon, D.; Dutcher, D.; Raymond, T. Characterization of *Halyomorpha halys* (brown marmorated stink bug) biogenic volatile organic compound emissions and their role in secondary organic aerosol formation. *J. Air Waste Manag. Assoc.* **2013**, *63*, 1264–1269. [CrossRef] [PubMed]

10. Fiola, J.A. Brown Marmorated Stink Bug (BMSB). Part 3—Fruit Damage and Juice/Wine Taint. In *Timely Viticulture*; University of Maryland Extension Publication: College Park, MD, USA, 2011.

11. Mohekar, P.; Lapis, T.J.; Wiman, N.G.; Lim, J.; Tomasino, E. Brown Marmorated Stink Bug Taint in Pinot noir: Detection and Consumer Rejection Thresholds of trans-2-Decenal. *Am. J. Enol. Vitic.* **2017**, *68*, 120–126. [CrossRef]

12. Mohekar, P.; Osborne, J.; Wiman, N.G.; Walton, V.; Tomasino, E. Influence of Winemaking Processing Steps on the Amounts of (*E*)-2-Decenal and Tridecane as off-Odorants Caused by Brown Marmorated Stink Bug (*Halyomorpha halys*). *J. Agric. Food Chem.* 2017. [CrossRef] [PubMed]

13. Fudge, A.L.; Schiettecatte, M.; Ristic, R.; Hayasaka, Y.; Wilkinson, K.L. Amelioration of smoke taint in wine by treatment with commercial fining agents. *Aust. J. Grape Wine Res.* **2012**, *18*, 302–307. [CrossRef]

14. Jackson, R.S. *Wine Science: Principles and Applications*; Academic Press: Cambridge, MA, USA, 2008; ISBN 978-0-08-056874-4.

15. Pickering, G.; Lin, J.; Reynolds, A.; Soleas, G.; Riesen, R. The evaluation of remedial treatments for wine affected by Harmonia axyridis. *Int. J. Food Sci. Technol.* **2006**, *41*, 77–86. [CrossRef]

16. Cosme, F.; Capão, I.; Filipe-Ribeiro, L.; Bennett, R.N.; Mendes-Faia, A. Evaluating potential alternatives to potassium caseinate for white wine fining: Effects on physicochemical and sensory characteristics. *LWT Food Sci. Technol.* **2012**, *46*, 382–387. [CrossRef]

17. Threlfall, R.T.; Morris, J.R.; Mauromoustakos, A. Effects of fining agents on trans-resveratrol concentration in wine. *Aust. J. Grape Wine Res.* **1999**, *5*, 22–26. [CrossRef]

18. Braga, A.; Cosme, F.; Ricardo-da-Silva, J.M.; Laureano, O. Gelatine, Casein, and Potassium Caseinate as distinct wine fining agents; different effects on colour, phenolic compounds and sensory characteristics. *J. Int. Sci. Vigne Vin* **2007**, *41*, 203–214. [CrossRef]

19. Fudge, A.L.; Ristic, R.; Wollan, D.; Wilkinson, K.L. Amelioration of smoke taint in wine by reverse osmosis and solid phase adsorption. *Aust. J. Grape Wine Res.* **2011**, *17*, S41–S48. [CrossRef]

20. Ugarte, P.; Agosin, E.; Bordeu, E.; Villalobos, J.I. Reduction of 4-Ethylphenol and 4-Ethylguaiacol Concentration in Red Wines Using Reverse Osmosis and Adsorption. *Am. J. Enol. Vitic.* **2005**, *56*, 30–36.

21. Ebeler, S.E. Analytical Chemistry: Unlocking the Secrets of Wine Flavor. *Food Rev. Int.* **2001**, *17*, 45–64. [CrossRef]

22. Styger, G.; Prior, B.; Bauer, F.F. Wine flavor and aroma. *J. Ind. Microbiol. Biotechnol.* **2011**. [CrossRef] [PubMed]

23. Jackson, R.S. *Wine Tasting: A Professional Handbook*; Academic Press: Cambridge, MA, USA, 2009; ISBN 978-0-08-092109-9.

24. Lawless, H.T. A simple alternative analysis for the threshold data determined by ascending forced-choice methods of limits. *J. Sens. Stud.* **2010**, *25*, 332–346. [CrossRef]

25. Heymann, H.; Noble, A.C. Comparison of Canonical Variate and Principal Component Analyses of Wine Descriptive Analysis Data. *J. Food Sci.* **1989**, *54*, 1355–1358. [CrossRef]

26. Peltier, C.; Visalli, M.; Schlich, P. Comparison of Canonical Variate Analysis and Principal Component Analysis on 422 descriptive sensory studies. *Food Qual. Prefer.* **2015**, *40*, 326–333. [CrossRef]

27. Bamforth, C.W. *The Oxford Handbook of Food Fermentations*; Oxford University Press: Oxford, UK, 2014.

28. Reynolds, A. *Managing Wine Quality: Oenology and Wine Quality*; Elsevier: Amsterdam, The Netherlands, 2010; ISBN 978-1-84569-998-7.

29. Sanborn, M.; Edwards, C.G.; Ross, C.F. Impact of Fining on Chemical and Sensory Properties of Washington State Chardonnay and Gewürztraminer Wines. *Am. J. Enol. Vitic.* **2010**, *61*, 31–41.

30. Ugliano, M. Oxygen Contribution to Wine Aroma Evolution during Bottle Aging. *J. Agric. Food Chem.* **2013**, *61*, 6125–6136. [CrossRef] [PubMed]

31. Barker, R.; Gracey, D.; Irwin, A.; Pipasts, P.; Leishka, E. Liberation of staling aldehydes during storage of beer. *J. Inst. Brew.* **1983**, *89*, 411–415. [CrossRef]

32. Chatonnet, P.; Dubourdieu, D. Identification of substances responsible for the "Sawdust" aroma in oak wood. *J. Sci. Food Agric.* **1998**, *76*, 179–188. [CrossRef]

33. Spillman, P.J.; Pollnitz, A.P.; Liacopoulos, D.; Pardon, K.H.; Sefton, M.A. Formation and Degradation of Furfuryl Alcohol, 5-Methylfurfuryl Alcohol, Vanillyl Alcohol, and Their Ethyl Ethers in Barrel-Aged Wines. *J. Agric. Food Chem.* **1998**, *46*, 657–663. [CrossRef] [PubMed]

34. Moreno, J.A.; Zea, L.; Moyano, L.; Medina, M. Aroma compounds as markers of the changes in sherry wines subjected to biological ageing. *Food Control* **2005**, *16*, 333–338. [CrossRef]
35. Culleré, L.; Cacho, J.; Ferreira, V. An Assessment of the Role Played by Some Oxidation-Related Aldehydes in Wine Aroma. *J. Agric. Food Chem.* **2007**, *55*, 876–881. [CrossRef] [PubMed]

Article

Novel Method for the Identification of the Variety of Grape Using Their Capability to Form Gold Nanoparticles

Silvia Rodriguez [1], Beatriz de Lamo [1], Celia García-Hernández [1], Cristina García-Cabezón [2] and Maria Luz Rodríguez-Méndez [1,*]

1   GroupUVaSens Department Inorganic Chemistry, Engineers School, Universidad de Valladolid, 47011 Valladolid, Spain; srodriguez_ross@hotmail.com (S.R.); beatriz.de.lamo.santamaria@gmail.com (B.d.L.); Celiagarciahernandez@gmail.com (C.G.-H.)
2   Materials Science Group, Engineers School, Universidad de Valladolid, 47011 Valladolid, Spain; anacrigar@gmail.com
*   Correspondence: mluz@eii.uva.es; Tel.: +34-983-423-540; Fax: +34-983-423-310

Received: 30 December 2017; Accepted: 19 March 2018; Published: 23 March 2018

Abstract: Gold nanoparticles (AuNPs) have been obtained using musts (freshly prepared grape juices where solid peels and seeds have been removed) as the reducing and capping agent. Transmission Electron Microscope images show that the formed AuNPs are spherical and their size increases with the amount of must used. The size of the AuNPs increases with the Total Polyphenol Index (TPI) of the variety of grape. The kinetics of the reaction monitored using UV-Vis shows that the reaction rates are related to the chemical composition of the musts and specifically to the phenols that can act as reducing and capping agents during the synthesis process. Since the particular composition of each must produces AuNPs of different sizes and at different rates, color changes can be used to discriminate the variety of grape. This new technology can be used to avoid fraud.

Keywords: gold nanoparticles; must; grapes

1. Introduction

The synthesis of gold nanoparticles is an active area of research, and a variety of techniques are currently available [1]. One of the most popular methods is the reduction of gold salts with an appropriate reducing agent, usually citrate. In recent years, green synthesis of gold nanoparticles (AuNPs) is gaining interest. In green synthesis, biomolecules (chitosan, polysaccharides, proteins, phenols, etc.) [2–7] and plant extracts (such as alfalfa, oats, coffee, onion, pear, banana, lemon grass extract, etc.) [8–11] are used as both the reducing agent and the stabilizer. AuNPs exhibit excellent physical, chemical and biological properties which are directly related to their size, shape and surface structure. For instance, AuNPs show a strong surface plasmon resonance (SPR) band in the visible region at ~520 nm [12,13] that is absent in bulk gold. The formation of gold nanoparticles can be detected by observing the color change of the solution. Moreover, the position, intensity and band-width of the SPR peaks rely on the nanoparticle size and shape [14,15]. Anions absorbed on the surface of nanoparticles stabilize the AuNPs avoiding aggregation. Different anions can be used to modulate the size and the color of the AuNPs. Size and shape dependent electrochemical and optical properties can be used for a wide range of applications, including chemical sensors for the assessment of the antioxidant capacity of foods [16–22]. According to these previous experiments, the optical properties of gold nanoparticles could be used to develop novel tools for the characterization of foods based on their antioxidant characteristics. Musts are extremely complex mixtures formed by more than 300 components including among many others ions, sugars, organic acids and a variety of antioxidants. Phenols are one of the most important classes of antioxidants. The synthesis of gold nanoparticles with phytochemicals (mainly phenols) present in grape seeds and peels has been described [23]. Since different varieties of grapes contain different concentrations of polyphenols and capping agents, one could expect that the type of grape strongly affects the shape and size of the NPs. On this basis, the shape or size on AuNPs could be used by official agencies to detect frauds relative to the use of grapes of a different variety than the stated in the label.

The objective of this work was to evaluate the possibility of discriminating musts produced from grapes of different varieties by means of their capability to obtain AuNPs and to analyze the effect of the grape variety in the size and shape of the nanoparticles. For this purpose, musts obtained from eight different varieties of grapes—Cabernet Sauvignon (C), Garnacha (G), Juan García (JG), Mencía Regadio (MR), Mencía Secano (MS), Prieto Picudo (PP), Rufete (R), Tempranillo (T)—were used as reducing and capping agents. The effect of the total polyphenolic content was discussed.

2. Materials and Methods

Chloroauric acid ($HAuCl_4 \cdot 3H_2O$) was purchased from Sigma-Aldrich (St. Louis, MO, USA) and was used as received without any further purification. The deionized water used in preparation was obtained from Millipore Direct Q TM (resistivity of 18.2 MΩ). Red grapes of eight different Spanish varieties were used to obtain musts. The varieties included in the study were: Cabernet Sauvignon (C), Garnacha (G), Juan García (JG), Mencía Regadío (MR), Mencía Secano (MS), Prieto Picudo (PP), Rufete (R) and Tempranillo (T). They were provided by the oenological station Castilla Leon and several wineries. Musts were prepared following a method developed by the Instituto Tecnologico y Agrario de Castilla y León (ITACyl) (Valladolid, Spain). 100 berries of the given variety of grape were introduced in a plastic bag and crushed for 1 min. Then, peels and seeds were separated by centrifugation (10 min at 5000 rpm). Total Polyphenol Index (TPI) was measured using international standard methods by measuring the Absorbance at 280 nm [24]. These results are collected in Table 1 along with other parameters of interest.

Table 1. TPIs measured by chemical methods.

Variety of Grape	TPI	pH
Cabernet Sauvignon	14	3.17
Garnacha	17	3.17
Juan García	24	3.39
Mencía Regadío	19	3.96
Mencía Secano	19	3.93
Prieto Picudo	26	3.35
Rufete	27	3.37
Tempranillo	24	3.30

AuNPs were obtained using musts as the reducing and the capping agent by mixing a certain volume of the corresponding must with a certain volume of the chloroauric 0.01 M. The proportions used are shown in Table 2. In this manner, 9 samples of each grape variety were prepared by varying the gold/must between 9:1 and 1:9, while keeping all other concentrations constant. All experiments were carried out at ambient temperature and atmospheric pressure.

Table 2. Volumes of Au^{3+} and musts employed for the synthesis of AuNPs (mL). Dilutions are denoted as M1 to M9.

ID	M1	M2	M3	M4	M5	M6	M7	M8	M9
Au^{3+}	9	8	7	6	5	4	3	2	1
Must	1	2	3	4	5	6	7	8	9

UV-Vis Spectra of AuNPs by bio-reduction of chloroauric acid in aqueous solution was recorded in a Shimadzu spectrophotometer (mod UV 2603) (Kyoto, Japan) operated at a resolution of 1 nm in absorption mode. UV-Vis spectra were acquired between 400 and 700 nm. Water was used as the reference and the blank for baseline subtraction. The bioreduction (formation of nanoparticles) of $HAuCl_4$ in the aqueous solution was monitored following the plasmon resonance absorption band of the reaction mixture for 5 h.

In order to investigate the effect of the gold to must ratio on the particle size, Transmission Electron Microscope (TEM) images were recorded on a high-resolution electron microscope (HRTEM: JEOL JEM 2200) (Tokyo, Japan) operating at an accelerating voltage of 200 kV. Sample images were processed using Image J image processing software (public software). The samples were prepared by drop casting the AuNPs solution on a carbon-coated copper TEM grid. Principal Component Analysis (PCA) was carried out using the software Matlab v5.3. (The Mathworks Inc., Natick, MA, USA).

3. Results

Colloidal AuNPs were obtained at room temperature by reducing Au^{3+} using musts obtained from red grapes obtained from eight different varieties of grapes as the reducing agent. The ability of musts to form nanoparticles was evaluated by monitoring the color changes of the solutions. In order to optimize the synthesis, musts were mixed with the gold salt in different proportions as indicated in Table 2.

As a first step, the formation of nanoparticles was followed visually. As observed in Figure 1, the color of the suspension changed upon addition of the gold salt from pale pink to dark purple which indicated the formation of AuNPs. The reaction started immediately and continued for several hours. It is worth noting that the extent of the reaction was dependent on the Au^{3+}:must ratio used, and also on the variety of grape; that is, on the chemical composition of the must (reduction capability and capping agents).

Figure 1. Response of mixtures of Au^{3+} salt and musts obtained from (**a**) Cabernet; (**b**) Garnacha; (**c**) Juan García; (**d**) Mencía Secano; (**e**) Mencía Regadío; (**f**) Prieto Picudo; (**g**) Rufete and (**h**) Tempranillo grapes. **Left column** corresponds to photographs taken 5 min after the mixture. **Right column** corresponds to photographs taken 3 h after the mixture. Ten tubes inside each picture correspond to increasing Au^{3+}:must ratios as indicated in Table 1.

The formation of AuNPs was monitored using UV-Vis spectroscopy (Figure 2). All spectra showed a broad absorption peak between 500 and 600 nm corresponding to the SPR band consistent with Mie theory [25,26], which establishes that the position of the SPR band is related to the size of AuNPs. In this case, the intensities and positions of the SPR bands were dependent on the nature of the grape used to obtain the musts. Since musts prepared from different varieties of grape have different compositions, the variety of colors obtained suggested that AuNPs could be used to analyze grapes.

For the smallest must volume (M1 where Au^{3+}:must = 9:1) the observed plasmon band was weak and appeared at ca. 580 nm. As the Au^{3+}:must ratio increased, a notable rise in the intensity of the SPR band was observed, confirming the increase in the number of nanoparticles obtained. Simultaneously, the position of the band shifted to lower wavelengths (at ca. 520 nm), suggesting growth of the nanoparticle size when increasing the reagent ratio. This is in good correlation with previous studies that have demonstrated that the size of the obtained nanoparticles depends on the concentration of the reducing agent (citrate, nitrite, etc.), and that this effect can be used to develop colorimetric probes [27,28]. A further observation from analysis of these spectra concerns the broadness of the peaks that are reflecting of a large dispersion of the size of the AuNPs obtained.

Similar trends were observed in most of the varieties analyzed. However, the spectral changes were smaller in those varieties of grapes with lower TPI in the whole range of 5 min–5 h. Figure 2 illustrates how the intensity of the peaks increase linearly with concentration.

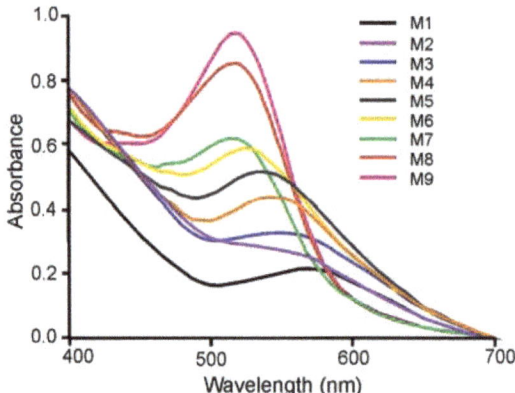

Figure 2. UV-Vis spectra of Au^{3+}:must Cabernet mixtures in proportions M1 to M9.

TEM images provided information about the morphology of the obtained AuNPs that confirms the information obtained from the UV-Vis absorption spectra. Figure 3 shows the images of the AuNPs obtained from Tempranillo must using different ratios Au^{3+}:must. The nanoparticles obtained showed a reasonable uniform size. When increasing the ratio Au^{3+}:must from M1 to M9, the size of the AuNPs increased from 5 to 60 nm.

In addition, the size and shape of the obtained nanoparticles changed with the type of must.

A closer look to the AuNPs revealed interesting features. In the case of Tempranillo grapes, the nanoparticles obtained were nearly spherical in shape, but, gold nanostructures with branched arms were also observed. This dual structure containing spherical nanoparticles and branched structures could indicate that two different growth mechanisms were involved. When increasing the Au^{3+}:must ratio from M1 to M9, the size of the AuNPs increased and the size of the branched arms grows simultaneously (Figure 4). The size and shape of the obtained nanoparticles changed with the type of must. When musts with low TPI were used as reducing agent, a larger number of AuNPs was obtained, and the shapes were not spherical. Instead, they were triangular or hexagonal. This is illustrated in Figure 5 for the AuNPs obtained using Garnacha grapes in ratio M1.

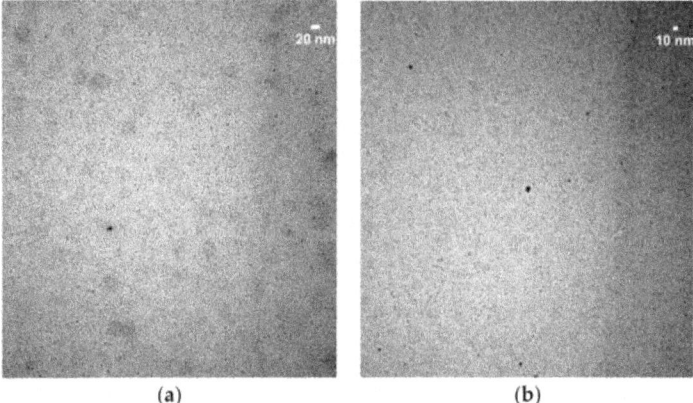

Figure 3. TEM images of the AuNPs obtained using Tempranillo must using (**a**) ratio M5; (**b**) ratio M9. Pictures taken at 2,000,000×; 80 K.

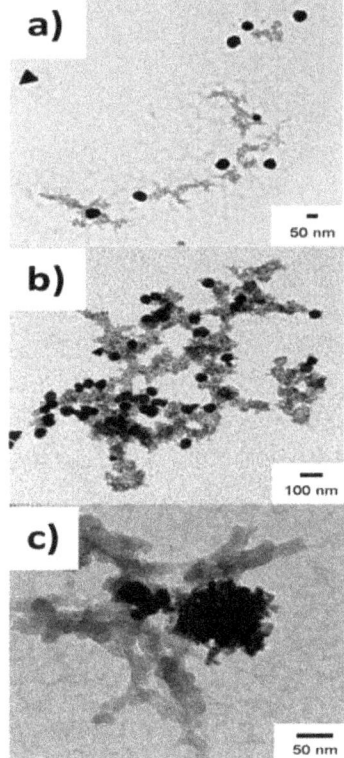

Figure 4. TEM images of the AuNps obtained using Tempranillo must using (**a**) M1; (**b**) M5 and (**c**) M9 ratios.

Figure 5. TEM images of the AuNps obtained using Garnacha must using Au^{3+}:must ratio 9:1.

As shown in the previous paragraphs, musts obtained from different varieties of grapes produced different amounts of AuNPs. Changes in color also indicate that some varieties of grape produce faster reactions than others. The kinetics of the reactions was studied using UV-Vis spectroscopy using three Au^{3+}:must ratios (M1, M5 and M9) for the different types of grapes. UV-Vis spectra were registered periodically after the addition of the corresponding must. Results are shown in Figure 6.

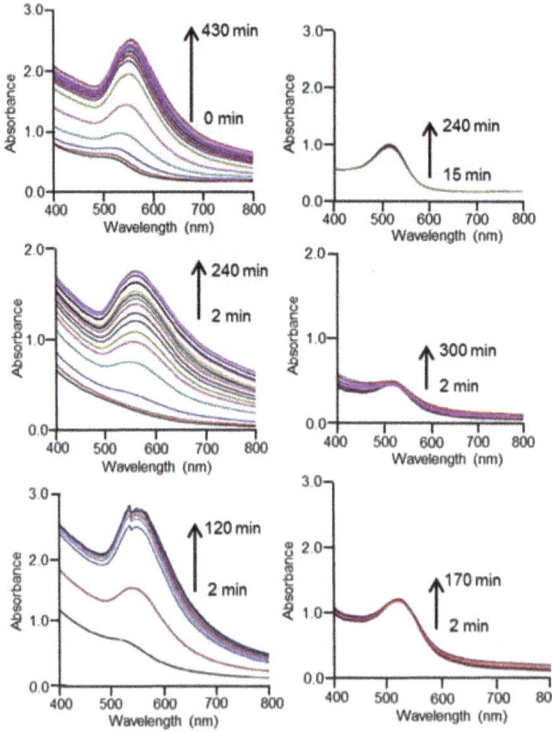

Figure 6. UV-Vis spectra of Au^{3+}:must mixtures in M1 (**left**) and M9 (**right**) using a must with low TPI (upper specta: Cabernet-Sauvignon), a must with medium TPI (middle spectra: Mecía Secano) and a must with high TPI (lower spectra: Tempranillo).

In all grapes studied, the absorbance increased progressively with time, confirming the formation of AuNPs. Then, a plateau was attained. The time elapsed before the constant value was reached was ca. 300 min when using varieties with low TPI such as Cabernet-Sauvignon. The time necessary to attain the plateau was ca. 180 min for varieties, such as Mancía Secano, with intermetadiate TPI values, and 50 min when varieties, such as Tempranillo, with high TPI were used. This result confirmed the influence of the antioxidant capacity in the AuNP formation.

Figure 7 illustrates the variation of the absorbance and the wavelength of the plasmon peak with time for the variety Tempranillo. Then, a plateau was reached. When AuNPs are formed using the proportion M1, the concentration of reducing compounds was small, and the formation rate of AuNPs was slow, as indicated by the slow changes in absorbance and wavelength observed. The rate of AuNP formation was faster using the proportion M5, where the formation of AuNPs was completed in a few seconds. In the case of the proportion M9, the formation rate was so fast that at the first measurement, the AuNPs had already been formed. The kinetics of the reaction analyzed by representing the slope of the change of absorbance with time (measured from 2 min after the addition until stabilization of the signal) confirmed the influence of the concentration of reactants in the formation rate. The rest of the musts showed similar trends, and reaction rates were directly related to the TPI values. Since the particular composition of each must produce AuNPs with different sizes and at different rates, color changes can be used to discriminate the variety of grape.

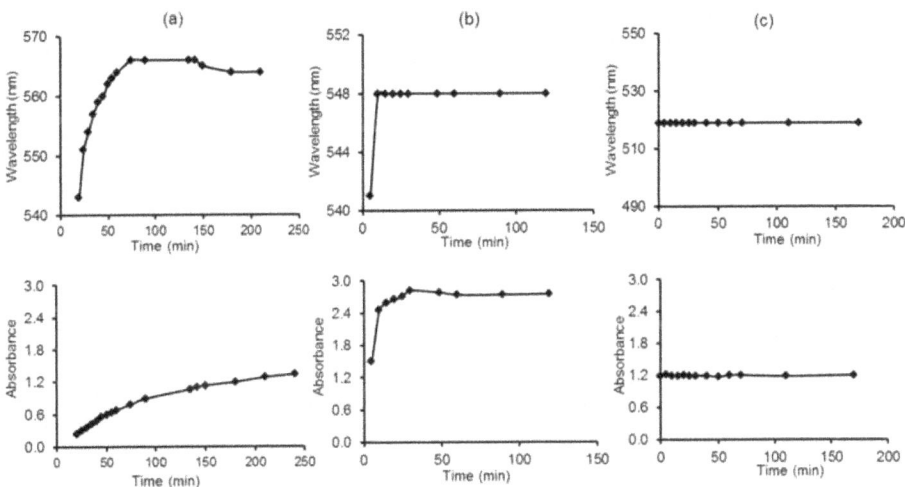

Figure 7. Variation of the wavelength and the absorbance with time during AuNPs formation using Tempranillo must (**a**) M1; (**b**) M5 (**c**) M9.

4. Discussion and Conclusions

Must-stabilized AuNPs of different diameters were prepared using grape juice. The influence of the proportion of grape juice and the variety of grape was analyzed. UV-Vis and TEM images show that when increasing the concentration of reactants, the particle size and the reaction rate increased simultaneously. Because musts prepared from different varieties of grape possess distinct chemical compositions, AuNPs of different sizes are obtained from different grapes. The particular reactivity shown by different variety of grapes produce AuNPs with different colors and with distinct kinetics. These observations can be used to discriminate grapes of different varieties and to analyze their phenolic content.

Acknowledgments: Financial support by MINECO and FEDER (AGL2015-67482-R) and the Junta de Castilla y León (VA011U16) is gratefully acknowledged. C.G.-H thanks for the grant of JCYL (BOCYL-D-24112015-9). S.R. would like to acknowledge EMA2 EURICA program for the scholarship granted.

Author Contributions: S.R., B.d.L. and C.G.-H. carried out the experiments. C.G.-C. interpreted the results. M.L.R.-M. interpreted the results, obtained the funds and wrote the paper.

Conflicts of Interest: Authors declare no conflicts of interest or state. The founding sponsors had no role in the design of the study; in the collection, analyses, or interpretation of data; in the writing of the manuscript, and in the decision to publish the results.

References

1. Zhao, P.; Li, N.; Astruc, D. State of the art in gold nanoparticle synthesis. *Coord. Chem. Rev.* **2013**, *257*, 638–665. [CrossRef]
2. Anuradha, S.; Anand, V.R.; Hemanth, K. Surface modification of chitosan for selective surface–protein interaction. *Carbohydr. Polym.* **2006**, *66*, 321–332. [CrossRef]
3. Bhumkar, D.R.; Joshi, H.M.; Sastry, M.; Pokharkar, A.V. Chitosan Reduced Gold Nanoparticles as Novel Carriers for Transmucosal Delivery of Insulin. *Pharmacol. Res.* **2007**, *24*, 1415–1426. [CrossRef] [PubMed]
4. Malathi, S.; Balakumaran, M.D.; Kalaichelvan, P.T.; Balasubramanian, S. Green synthesis of gold nanoparticles for controlled delivery. *Adv. Mater. Lett.* **2013**, *4*, 933–940. [CrossRef]
5. Schulz, A.; Wang, H.; Van Rijin, P.; Boker, A. Synthetic inorganic materials by mimicking biomineralization processes using native and non-native protein functions. *J. Mater. Chem.* **2011**, *21*, 18903. [CrossRef]
6. Arakaki, A.; Shimizu, K.; Oda, M.; Sakamoto, T.; Nishimura, T.; Kato, T. Biomineralization-inspired synthesis of functional organic/inorganic hybrid materials: Organic molecular control of self-organization of hybrids. *Org. Biomol. Chem.* **2015**, *13*, 974–989. [CrossRef] [PubMed]
7. Galloway, J.M.; Staniland, S.S. Protein and peptide biotemplated metal and metal oxide nanoparticles and their patterning onto surfaces. *J. Mater. Chem.* **2012**, *22*, 12423–12434. [CrossRef]
8. Gardea-Torresdey, J.L.; Parsons, J.G.; Gomez, E.; Peralta-Videa, J.; Troiani, H.E.; Santiago, P.; Jose-Yacaman, M. Formation and Growth of Au Nanoparticles inside Live Alfalfa Plants. *Nano Lett.* **2002**, *2*, 397–401. [CrossRef]
9. Bankar, A. Banana peel extract mediated synthesis of gold nanoparticles. *Coll. Surf. B Biointerfaces* **2010**, *80*, 45–50. [CrossRef] [PubMed]
10. Huang, J.; Li, Q.; Sun, D.; Lu, Y.; Su, Y.; Yang, X.; Wang, H.; Wang, Y.; Shao, W.; He, N.; et al. Biosynthesis of Silver and Gold Nanoparticles by Novel Sundried Cinnamon camphora Leaf. *Nanotechnology* **2007**, *18*, 105104. [CrossRef]
11. Shankar, S.S.; Rai, A.; Ankamwar, B.; Singh, A.; Ahmad, A.; Sastry, M. Biological Synthesis of Triangular Gold Nanoprisms. *Nat. Mater.* **2004**, *3*, 482–488. [CrossRef] [PubMed]
12. Haiss, W.; Thanh, N.T.; Aveyard, J.; Fernig, A.D. Determination of Size and Concentration of Gold Nanoparticles from UV-Vis Spectra. *Anal. Chem.* **2007**, *79*, 4215–4221. [CrossRef] [PubMed]
13. Sharma, A.; Singh, B.P.; Gathania, A.K. Synthesis and characterization of dodecanothiol-stabilized gold nanoparticles. *Indian J. Pure Appl. Phys.* **2014**, *52*, 93–100.
14. Olson, J.; Dominguez-Medina, S.; Hoggard, A.; Wang, L.; Chang, W.W.; Link, S. Optical Characterization of Single Plasmonic Nanoparticles. *Chem. Soc. Rev.* **2015**, *44*, 40–57. [CrossRef] [PubMed]
15. Yang, X.; Yang, M.; Pang, B.; Vara, M.; Xia, Y. Gold nanomaterials at work in biomedicine. *Chem. Rev.* **2015**, *115*, 10410–10488. [CrossRef] [PubMed]
16. Cobley, C.M.; Chen, J.; Cho, E.C.; Wang, L.V.; Xia, Y. Gold nanostructures: A class of multifunctional materials for biomedical applications. *Chem. Soc. Rev.* **2011**, *40*, 44–56. [CrossRef] [PubMed]
17. Kong, D.; Liu, L.; Song, S.; Suryoprabowo, S.; Li, A.; Kuang, H.; Wang, L.; Xu, Ch. A gold nanoparticle-based semi-quantitative and quantitative ultrasensitive paper sensor for the detection of twenty mycotoxins. *Nanoscale* **2016**, *8*, 5245–5253. [CrossRef] [PubMed]
18. Medina-Plaza, C.; García-Cabezón, C.; García-Hernández, C.; Bramorski, C.; Blanco-Val, Y.; Martín-Pedrosa, F.; Kawai, T.; de Saja, J.A.; Rodríguez-Méndez, M.L. Analysis of organic acids and phenols of interest in the wine industry using Langmuir-Blodgett films based on functionalized nanoparticles. *Anal. Chim. Acta* **2015**, *853*, 572–578. [CrossRef] [PubMed]

19. Medina-Plaza, C.; Furini, L.N.; Constantino, C.J.L.; de Saja, J.A.; Rodríguez-Mendez, M.L. Synergistic electrocatalytic effect of nanostructured mixed films formed by functionalised gold nanoparticles and bisphthalocyanines. *Anal. Chim. Acta* **2014**, *851*, 95–102. [CrossRef] [PubMed]
20. Yola, M.L.; Atar, N.A. Novel voltammetric sensor based on gold nanoparticles involved in p-aminothiophenol functionalized multi-walled carbon nanotubes: Application to the simultaneous determination of quercetin and rutin. *Electrochim. Acta* **2014**, *119*, 24–31. [CrossRef]
21. Vilela, D.; González, M.C.; Escarpa, A. Nanoparticles as analytical tools for in-vitro antioxidant-capacity assessment and beyond. *Trends Anal. Chem.* **2015**, *64*, 1–16. [CrossRef]
22. Vilela, D.; González, M.C.; Escarpa, A. Gold-nanosphere formation using food sample endogenous polyphenols for in-vitro assessment of antioxidant capacity. *Anal. Bioanal. Chem.* **2012**, *404*, 341–349. [CrossRef] [PubMed]
23. Amarnath, K.; Mathew, N.L.; Nellore, J.; Siddarth, C.R.V.; Kumar, J. Facile synthesis of biocompatible gold nanoparticles from Vitis vinifera and its cellular internalization against HBL-100 cells. *Cancer Nanotechnol.* **2011**, *2*, 121–132. [CrossRef] [PubMed]
24. International Organisation of Vine and Wine (OIV). Compendium of International Methods of Analysis of Wines and Musts. In *Bulletin de L'organisation Internationale de la Vigne et du Vin*; OIV: Paris, France, 2013.
25. Klar, T.; Perner, M.; Grosse, S.; Von Plessen, G.; Spirkl, W.; Feldmann, J. Surface-Plasmon Resonances in Single Metallic Nanoparticles. *Phys. Rev. Lett.* **1998**, *80*, 4249–4252. [CrossRef]
26. Kelly, K.L.; Coronado, E.; Zhao, L.L.; Schatz, G.C. The optical properties of metal nanoparticles: The influence of size, shape, and dielectric environment. *J. Phys. Chem. B* **2003**, *107*, 668–677. [CrossRef]
27. Nama, Y.S.; Noh, K.C.; Kim, N.K.; Lee, Y.; Park, H.K.; Lee, K.B. Sensitive and selective determination of NO2 ion in aqueous samples using modified gold nanoparticle as a colorimetric probe. *Talanta* **2014**, *125*, 153–158. [CrossRef] [PubMed]
28. Medina-Plaza, C.; Rodriguez-Mendez, M.L.; Sutter, P.; Tong, X.; Sutter, E. Nanoscale Au-In alloy-oxide core-shell particles as electrocatalysts for efficient hydroquinone detection. *J. Phys. Chem. C* **2015**, *119*, 25100–25107. [CrossRef]

MDPI

St. Alban-Anlage 66

4052 Basel

Switzerland

Tel. +41 61 683 77 34

Fax +41 61 302 89 18

www.mdpi.com

Beverages Editorial Office

E-mail: beverages@mdpi.com

www.mdpi.com/journal/beverages

www.ingramcontent.com/pod-product-compliance
Lightning Source LLC
Chambersburg PA
CBHW041144120626
46547CB00020B/3107